# ESSENCE

## o f

# EXISTENCE

*Brief story of matter and people*

PAUL ROMAN

❖

Promanq
book

ISBN 978-0-9918399-3-3

Library and Archives Canada Cataloguing in Publication

Roman, Paul, 1944-, author
   Essence of existence : brief story of matter and people
/ Paul Roman.

Includes bibliographical references.
ISBN 978-0-9918399-3-3 (pbk.)


   1. Cosmology.  2. Life--Origin.  3. Evolution.  4. Human
beings--Origin.  5. Social evolution.  I. Title.

B511.R65 2014      113      C2014-907733-5

Editor: Ann Westlake
Cover:
   Design: © Paul Roman
   Photography:
   1. Cygne Nebula © Shot by Subaru telescope of
      National Astronomical Observatory of Japan.  All
      rights reserved.  Used with permission.
   2. Australopithecine in countryside, photomontage.
      a/ Model of Australopithecine in American Museum
      of Natural History. Shot by Ryan Somma. Some
      rights reserved.
      b/ Alberta Badlands.  Shot by Paul Roman.Some
      rights reserved.
Illustrations: © Paul Roman
Published by PromanP Books

## ABOUT THE AUTHOR

Paul is a retired architect and now occasionally teaches downhill skiing. He writes on subjects of natural science and society development to clarify and expand his world perception and records it for interested adolescent youth and lay adults. Right now he is working on the sequel to this book. Paul escaped the communist tyranny of then Czechoslovakia in 1968 and has lived in Canada ever since.

## OTHER WORKS BY PAUL ROMAN

**Podstata Existence** - ISBN 978-80-7453-268-9
Vydala Nová Forma v měkké vazbě v roce 2012.
www.stahuj-knihy.cz/stahujknihy/eshop/21-1-literatura-faktu/0/5/540-Roman-Pavel-Podstata-Existence
Kniha stručně vysvětluje jak vznikl vesmír, život, člověk i společnost. To je úžasný příběh poutavě zde předložený pro zvídavou mládež i dospělé.

**Chalupářství v Jizerkách** - ISBN 978-0-9918399-1-9
Vydala PromanP kniha v měkké vazbě v roce 2014.
www.stahuj-knihy.cz/stahujknihy/eshop/11-1-HISTORIE/0/5/810-Roman-Pavel-Chaluparstvi-v-Jizerkach//description#anch1
Odezva dvou válek v osobních vzpomínkách.

Website: www3.telus.nct/p-roman/

TO:
My uncle Jára - in memoriam,
   Because without his nudging I probably
   wouldn't have started writing.

My grandchildren,
   Because, one day, they will find a good
   textbook in this story.


## ACKNOWLEDGEMENTS

My sincere thanks and gratitude go to Charles
Nemeth, Vladimir Matus, Jiří Pavlásek and
Lindsay Franklin who provided me with
substantive and editing comments diligently,
from the begining to the end, and selflessly, for I
probably won't have an opportunity and ability
to return the favour.

## CONTENTS

About the Author iii
Preface vi
*COSMOS* 10
  Age of darkness 10
  Stars 14
  Planet Earth 15
*LIFE* 26
  Organic chemistry 26
  Bay of life 27
  Math rules 28
  Crystallization of life 30
  Plan of life 32
  Directing orders 34
  Stillness, chaos, order 36
*EVOLUTION* 39
  Nothing is infallible 39
  Fitness landscape of evolution 45
  Sex 49
  Mathematics still rules 51
  Key turning points 52
*MAN* 71
  Compressed time 71
  Cradle of mankind 72
  Birth of technology 75
  Social evolution 77
*APPENDIX 1* 98
*APPENDIX 2* 103
*APPENDIX 3* 106

# PREFACE

This book describes, briefly and simply, basic knowledge covering a relatively broad area. I would have liked to find such a concise book myself when I had no time to collect information that, despite being very interesting to me, wasn't really necessary for my profession. As an architect, I belong to a type of people who typically tend to be generalists rather than specialists. Maybe that's why I strive to get to the essences of things quickly, and then describe them briefly.

Most of the knowledge contained in the following story comes neither from my formal education nor from work experience obtained in my three careers. Nearly everything originates from unstructured reading of various books – mostly popular science, conversations or other acoustic and visual sources.

A part of what follows perhaps comes from original thinking, but that is uncertain. The information stored in my brain can be divided into three categories. Firstly there is information from known sources. Then there is information that I know came from forgotten sources. Lastly, there is information about which I am not sure whether it came to me during occasional thinking or from long forgotten sources.

I dare to say that rational people have at least the two first categories of information in their brains. Most people likely have three categories of information, the same as I do. Probably, very few individuals possess the fourth category:

substantial and original knowledge derived from their own thoughts. I have neither the illusion nor do I feel any need to belong among them. Then there must be a number of people who pride themselves on having original thoughts that are actually not original but originate from forgotten or even subconsciously or deliberately unacknowledged sources. I wouldn't like to belong to this group. Hopefully, I have managed to put a few pieces of information together in a novel way to make a new point of view.

Under these circumstances, it is impossible to provide references to all sources of information. I will provide references, where able, to enable readers to explore my sources of information.

Personally I am content with only a glimpse of understanding the fundamental principles of the universe and inorganic and organic matter within it. I put more emphasis on understanding the essence of humanity and our civilization with its broad and complicated contents. My aim is to provide the beginning of a journey to more detailed knowledge of these subjects.

It took me a while to choose a form of narration. Finally, I chose an undefined narrator who talks about phenomena and objects in the present tense. I hope that this form of narration will better draw readers into a distant past and into environments that are unfamiliar to them, or objectively unknowable. But it is not necessary for the narrator to stay chronologically in the past in a rigorous way. Sometimes it is better to jump back and forth in time.

❖

You are now at the beginning of a journey that will cover about 14 billion years of nature's history. When you get to the end of the book, you will find a few simple questions. Please consider answering them. The author will be grateful for each response.

## COSMOS

AGE OF DARKNESS. With the beginning
of the universe we tackle a problem since an
absolute beginning cannot be quite grasped by the
human mind. The mind immediately asks the
question: What was there before the beginning?
If there was nothing, we cannot imagine this
"nothing". That's why it is perhaps better to
imagine that whatever was there at the beginning
has existed since infinity. But we are not able to
imagine infinity either.

Our story begins about 14 billion years ago. A
year is, however, still an utterly meaningless
term because time does not exist. The beginning
is indeed very inexplicit, as there are no witnesses.
It is a non-moment about which only hardly visible traces are left behind.

The mass of matter that will ever exist is now
compressed into a tiny bead similar in size to a
poppy seed, which will not exist for another
10 billion years. The matter is extremely dense and
hot; it has no history, but has unlimited potential.
It is very tight because space does not exist yet.
But it is not alone. In addition to this matter, there
is also energy, which compresses the matter into
a small bead. Then there is also information,
which gives the matter its unlimited potential;
the matter would not know what to do without it.

It is possible, or even probable, that matter
doesn't exist either, since it may have somehow
evaporated into energy due to the enormous heat
in the compressed bead. But the moment this pre-

existing bead ceases to exist, the evaporated energy will start to convert into matter in the very first instance of newly-emerged time. Therefore, at least potentially, matter does exist even in the pre-existing bead.

Matter can be imagined, rather inaccurately, as something touchable. Energy can be imagined as a force that can push matter around. To imagine information is not easy. We can call it a set of rules, laws of nature, Mother Nature, goddess or God, or anything else. In this story, we will call information simply Nature. Whether it is a personalized natural force, a supernatural force, or just an attribute of matter, information is not easily discernible; it's hugely complicated and admirable indeed.

For the time being, matter is inactive. Then suddenly Nature brings about a big bang, or explosion, that is accompanied by an unheard noise, or perhaps a peculiar silence. Tightness is instantly replaced by a huge and expanding space sparsely occupied by extremely small particles of matter. Also, time starts ticking, and it heads infallibly to its unimaginable end, or possibly to eternity. As the small pieces of matter dash in all directions, cosmic space expands staggeringly fast each second. But one second doesn't mean much for the cosmos. Even a million years is not a very long time in the cosmic scale.

Flying matter doesn't look very interesting. It forms thin and chaotic groups of extremely small quarks, protons, neutrons, electrons and other subatomic particles. As they fly, particles collide with each other, and sometimes various forces bind them together so that they form – exactly as they were ordered by Nature – atomic nuclei. The nuclei can then fall apart again due to collisions with other pieces of matter. As long as forces and matter are in balance, the particles of matter stay together. When the external force is greater than the binding force, particles fall apart again.

Atomic nuclei attract electrons that are now circling around them. The bigger the nucleus the more electrons can revolve around it. Atomic nuclei with an electron or electrons form atoms of chemical elements. The atoms are much bigger than subatomic particles, yet they are still invisible. For now, only two chemical elements have sprung up: hydrogen and helium. Hydrogen has only one electron circling around its nucleus of only one proton. Helium has two electrons circling around its nucleus consisting of two protons. Similarly, as some forces keep electrons close to nuclei, other forces keep atoms of one chemical element close to each other. Those forces that act upon atoms of hydrogen and helium are relatively weak, so their atoms don't cluster too tightly and move around freely. Materials of this nature will eventually be called gases.

No chemical elements other than hydrogen and helium exist now, because it is not hot enough in the universe. The enormous heat of the pre-existing bead actually dissipated quickly into the ever enlarging and ever colder cosmic space. Light doesn't exist yet, that's why this period will eventually be called The Age of Darkness. So the atoms of hydrogen and helium tear through the freezing darkness together with many subatomic particles.

One of the rules of Nature is the rule of inertia that compels any object to keep moving through space in the same direction and at the same speed, provided that no external force changes its speed and direction. Another rule is that all pieces of matter are attracted to each other. This attracting force will later be called gravity. The distribution of matter in space is not even but lumpy. Here and there, under the influence of gravity, matter gathers in large wisps or nebulae. The larger the nebulae, the more they attract ambient matter because the force of gravity grows with the mass of matter. The densest nebulae are eventually compressed by gravity into large spheres. The spheres are huge, much bigger than our sun will be, much much later.

As the huge spheres continue to grow, the compressed particles of matter randomly vibrate or jiggle with increasing speed. The result of this jiggling is friction between atoms, which is essentially heat energy of matter, so eventually hydrogen in one of the compressed spheres reaches such a high temperature that it starts to glow. Thus, the first cosmic star is born. As if burning, the glowing star radiates light into the freezing universe. Then, like individual matches, other dense cosmic spheres start glowing in succession, ending The Age of Darkness nearly 100 million years after the big bang. The stars send out glowing light with staggering speed to all corners of the universe.

If light moves faster than matter expanding through space from the location of the big bang, then the whole universe will be flooded by light; if it doesn't, then an impenetrable envelope of darkness will remain at the edge of the universe. We'd rather not ask what's behind this envelope.

STARS. A period of 100 million years is an unimaginably long period of time for us, but for Nature it is just a flash. The first stars are now immense laboratories that Nature needs for further cosmic evolution.

Progressively heavier and heavier atomic nuclei are made by fusing subatomic particles together in glowing hydrogen-fuelled furnaces of countless stars. The heavier the nucleus, the more electrons can circle around it. The nuclei and electrons together form atoms of chemical elements. Many atoms of new chemical elements are gradually made. They will be called lithium, beryllium, carbon, oxygen silicon, iron and many others. In comparison with The Age of Darkness, the universe is starting to get complicated.

After another 3 million years, a new unstable situation develops, and stars begin exploding, one after another. The newly-shattered matter again speeds through cosmic space from locations where exploded stars used to be. But now it contains way more interesting matter than before. It consists of many different atoms. They now cluster again and form new spheres. The spheres are compressed by gravity, warm-up and eventually ignite as stars of a new age. One of them, our own sun, lights up about 4 to 8 billion years after the big bang.

In general, new-age stars are much smaller and denser than the stars that gave origin to light at the end of The Age of Darkness. They are composed of more diverse matter and will last much longer. Some of them will glow for 20 billion years, perhaps even longer.

*EARTH.* The denser and bigger a lump of the matter is, the more it attracts other matter. If a piece of matter has extremely small mass, its gravity is negligible. Two pieces of matter that get close will affect each other's trajectory and speed as the force of gravity affects them. Due to inertia, faster moving particles bend their trajectory less than slower particles. The mass of the sun is so huge that many chunks of matter bend their trajectory so much that they collide with the sun instead of passing by and are absorbed by it. More distant pieces of matter swing around the sun and continue circling for perpetuity.

In this way, our sun slowly captures matter from the adjacent universe, which accumulates into thick nebulae at several locations in various distances from the sun. Gravity compresses the nebulae into dense spheres that become planets circling around the sun. The process of compression increases the temperature of the matter again; so early planets are spheres of hot gases, but they are much cooler than the stars. One of the planets is our Earth, born about 4 or 4.5 billion years before the present time.

As with the sun, Earth also captures matter from the universe that circled around it; the captured matter then slowly thickens into a small nebula and finally into our moon that will continue dancing in a circle around Earth in perpetuity; but the moon could have also formed in several different ways.

By size, the moon and Earth are not even remotely similar to the sun and other stars. They are a lot smaller by far. We can't imagine how much smaller they are because our mind can comprehend relative size only of things that surround us every day.

To help our imagination, we have to picture compressed imaginary celestial bodies. By our mind's eye, let's try to uniformly compress all cosmic bodies to make the moon as large as a mustard seed. Then Earth will be as large as a pea, planet Uranus will be as a plum. Jupiter, the biggest planet of our solar system will be as a large orange. The sun will be approximately the size of a sheep. Pollux, a medium-sized star, will be already as large as a pick-up truck. The gigantic star Rigel in the constellation Orion will be as large as a hill protruding 50 metres above the surrounding countryside. The super gigantic star Antares will then be as big as a mountain towering 700 metres over a deep valley.

Right from its birth, the hot Earth radiates heat into the freezing cosmic space and thus begins to cool down. Different atoms of chemical elements circulate within the hot gases of the sphere Earth. They often collide with each other and, as long as all of the external and internal forces are in equilibrium, the atoms remain bound into larger, stable groups of atoms. These groups will be called molecules. However, not just any group of atoms can form molecules; only exactly chosen groups of atoms can do so.

Nature does not have to assist all atoms in binding together; that would be too much work. It only established the rules according to which the binding must be governed. That's why molecules spring up by themselves, always in accordance with the laws determined ahead of time.

These laws are immensely complicated, and we don't need to understand them. For illustration, we can just imagine atoms as small balls with little protrusions and holes, which have many different shapes. Some protrusions fit perfectly into some holes, as keys to keyholes. When atoms collide in such a way that a key fits a lock, a chemical bond results and atoms stay together as a molecule – unless some external force knocks them apart again. The atoms on the periphery of these molecules have theirs keys and keyholes exposed to the surrounding space, so even molecules can be bound together if a key fits a lock. Or we can imagine atoms and molecules as space ships that can join if they have landing docs that fit together.

Sometimes the docks of two approaching molecules are already occupied, so in order for a landing to occur, some atoms of one or both molecules must be first released or knocked off their docks. That's how chemical reactions ensue. Atoms that are knocked off and released during the landing sometime join together, so that one landing procedure results in several different molecules.

Many molecules can arise only at high temperature because heat compels atoms to oscillate or jiggle quickly and randomly. Light atoms and molecules jiggle more than heavy atoms and molecules. This molecular motion brings about many collisions and opportunities for various bonding. The likelihood of new chemical bonds increases with a rise in temperature.

Earth is extremely hot, so there is no lack of opportunities for chemical bonding. Over 100 kinds of atoms oscillate around each other and enable formations of many new molecules – chemical substances. One of them, for instance, is the molecule of carbon dioxide – $CO_2$.

Two atoms of hydrogen often approach one atom of oxygen within the never-ceasing dance of atoms. These atoms befit each other well, so they join as one molecule of water – actually water vapour. Molecules of water vapour are lighter than molecules of $CO_2$ and many other gases, so they float up close to the surface of the red-hot gaseous sphere Earth. As Earth cools down, bigger and heavier molecules oscillate less and less, spaces between molecules shrink, and clouds of gases become thicker. This cooling process continues, molecules sluggishly mill around until, quite suddenly, one very special effect of Nature comes into existence for the first time – the change of state of matter: gases composed of heavy molecules now liquefy into thick liquids. Molecules of liquids are closer together because they move around with slower speed, and a casual observer would easily guess, by looking at visible surface currents, that molecules of liquids are still moving around. But clearly there is no observer on Earth yet. Liquids are heavier than gases, so the early thick, porridge-like liquids sink towards the centre of Earth, which is now divided into the liquid centre (magma) – and its gaseous jacket (the atmosphere). More and more gases now liquefy until finally the atmosphere is only a very thin layer on the surface of magma.

The mushy Earth surface continues to lose heat into the cosmic space, and it's just about time for another significant effect of Nature. With lower temperatures all molecules move around with slower speed until, quite suddenly, a group of the heaviest molecules freezes into a solid state. The first solid matter begins to float as a crust on the magma of planet Earth. An observer could get a sense that all movement of molecules stopped. Molecules at the still surface of solid matter look like an impenetrable mass. It appears that molecules are pressed together so densely that they touch each other in rigid immobility. But this appearance is deceiving. In-between molecules of solid matter is so much space that small subatomic particles, permanently flying through the cosmic space, penetrate and fly through the solid matter with ease. If the matter is thin, a speeding particle probably doesn't even touch a single molecule.

With a little additional cooling, another solid matter now appears in Earth. At specific level of temperature and pressure, molecules of iron start to solidify. Iron is hefty, so it sinks to the center of Earth.

As Earth circles around the sun, it also spins around its axis. Lighter crust and liquid magma spins little faster than heavy, solid iron core; this is the reason why magnetic field now starts emanating from poles and envelopes the whole Earth. Some particles of matter, which keep hurtling through cosmic space, are deflected by the magnetic field. Without the magnetic field shield, the atmosphere would be slowly ripped off from Earth.

Just as we cannot imagine unusually large cosmic bodies, likewise we can't imagine unusually small things. We can only clearly imagine whatever frequently surrounds us. For illustration how molecules, atoms and the spaces between them are small, in our mind's eye we uniformly enlarge everything in such a way as to make a common house fly as large as a city with half a million residents. A flea will then be as large as an airport. Large organic cells will be as large as a family house, and small cells will be as large as a refrigerator. Viruses will be as a small plum. Giant protein molecules will be as peas, and large organic molecules will be as poppy seeds.[1] Small molecules, atoms and spaces between them will be so small even in this grossly enlarged scale that we have no comparison of them to any object from our daily life. And the subatomic particles are so indescribably small that the space occupied by any atom with its nucleus and revolving electrons is well over 90 percent empty.

Seemingly motionless molecules in the solid crust of Earth's surface are still jiggling but much less than molecules of original liquids did. The slower movement of molecules results in a slower transfer of heat through the crust, so the Earth's surface is heated up (from the glowing depth of hot magma) less than before. As the surface radiates its heat into the universe through the atmosphere, the surface cools down rapidly. The atmosphere is now entirely separated from the hot magma by Earth's crust and it cools down even faster. The same process that condensed gases to liquids and formed magma at much higher temperature now influences water vapours. Some molecules of water vapour, as by the flick of a magic wand, now form droplets of water. The very first rain now falls on Earth's crust, lakes soon fill dips and hollows, and flowing water begins to scrape out stream beds of creeks, brooks and rivers, and it infallibly heads to the lowest points of the crust, where it fills seas and oceans.

---

[1] Horst Reinbothe; *Molekül, Mikrobe, Mensch*

Water isn't thick and mushy as the initial liquids of Earth's magma. Water is clear and very fluid. It is also easily influenced by another significant effect of Nature. That is evaporation. When heated up, water can evaporate – it becomes vapour again, its state changes from liquid to gas. The sun's rays heat up the surfaces of lakes and oceans from far above. Depths of oceans are often heated up by hot gases that tear up from magma through cracks in Earth's crust. The evaporation from Earth's emerging bodies of water is considerable, so rain torrents from clouds of gloomy skies are substantial.

The solid crust of Earth continues to lose heat into the freezing cosmic space. Magma liquids also lose heat into the cooler crust above. At a lower temperature, some parts of magma freeze into solid rock, thus making the crust thicker. But in some places, magma flows relatively quickly, bringing up heat from the deep. This way the crust heats up and another of Nature's remarkable effects comes into existence. It is melting or thawing. Molecules of solids sometimes jump back into a liquid state.

The cycling of chemical compounds between different states of matter on Earth is an ongoing process. When the temperature drops, it leads to liquefying and then freezing. When the temperature rises, it leads to melting and then evaporation. Various atoms and molecules change the states of matter at different temperatures. So, for example, pure hydrogen or nitrogen exist on Earth only in gaseous state because both liquefy and freeze at very low temperatures that don't exist on Earth, whereas iron or pure carbon exists in Earth's crust only in the solid state. Only a few materials exist on Earth in both gaseous and liquid state. Perhaps only water will exist in all three states on Earth at one time.

The thin skin of Earth's crust is not uniform or continuous. It is divided by cracks or faults, which are often covered by young oceans, and consist of separate plates forming Earth's crust. The plates float on hot magma and constantly move as they are pushed by magma's currents. However, this movement is exceedingly slow, entirely unnoticeable and measurable only by a few millimeters per year. Even so, as the edges of these plates slide by and grind against each other, Earth often quivers, and streams of fiery magma or lava sometimes tear up explosively through the cracks, and flow around hills and down to the bottoms of deep oceans.

Under relentless pressure of flowing magma, an edge of one plate sometimes slides under the edge of another plate. Those plates only overlap by small amounts initially, just a few millimeters after the first year. But one year isn't terribly important in the evolution of Earth. In geological time, even a period of a hundred million years is a relatively short time. However, after millions of years, the plates are overlapped by hundreds of kilometers. The surface of the upper plate rises slowly – only a few millimeters annually – as the lower plate floats on magma and is pushed up by buoyancy. This is the way mountains arise. Val-

ley bottoms rise less than peaks and ridges because streams have enough time to scrape out
crevices, galleys and wide valleys into those
slowly-rising mountains. After hundreds of millions of years, ridges and mountain peaks tower
thousands of metres over their surroundings.
Eroded fragments of rocks are carried away by
water and deposited at the bottoms of seas, and
are thus pushed down into the supportive magma.

Earth revolves around its axis which is close to
being perpendicular to the sun's rays. This axial
tilt from the perpendicular position slowly wob-
bles between $21.5°$ to $24.5°$. Half of the planet is
always in darkness; the other half is flooded by
light rays. The sun's warmth moves over the surface as the planet turns. One complete turn
equals one day. The two points where the imaginary axis of Earth intersects the planet's surface
will be called poles. The imaginary circle that is
equidistant from the poles will be known as the
equator. Here, sun rays beat onto Earth at right
angle, and they add heat, or thermal energy to
our planet. At the poles, sun rays fleetingly
touch the surface at a very low angle as they
bounce back into cosmic space. Overall, Earth
keeps losing warmth into the universe much faster at the poles than at the equator.

Originally, water occurred on Earth only as
vapor. Later on, it also appeared as liquid. But
now – as with magma, which recently changed
state from liquids to solid rock at Earth's surface
– water, now at much lower temperatures,

changes over to solid ice and snow at the poles and high mountains.

Earth's surface, veiled in a thin atmosphere, now consists of crust of rocks, pebbles and sand, water, and white snow and ice. The surface layer of snow and ice accelerates the cooling of Earth because white surfaces reflect sun rays well and do not absorb much heat. But the hot centre of Earth is well insulated from the freezing cosmic space by the crust and atmosphere, and so it will stay partially liquid, perhaps forever. Earth is achieving a thermal balance. It loses into space approximately the same amount of heat as it gains from the sun. Earth is ready for the most magical trick of Nature.

# LIFE

ORGANIC CHEMISTRY. The first molecules
created in the chemical laboratory Earth were
relatively small. They consisted of only a small
number of elements and a few atoms of each
element. Silicon dioxide ($SiO_2$) and oxide of
hydrogen ($H_2O$) serve as an example of small
molecules, and both belong to inorganic chemistry. The first one is a basic component of hard
granite stone; the second one is perhaps the best
known chemical compound – water. They are
unbelievably dissimilar materials, both created
from only 3 atoms of 2 elements. That is because
the rules of Nature don't allow most atoms to
form large molecules. They do not have enough
keys and keyholes, nor enough suitable landing
docks.

   But this certainly does not apply to carbon
atoms, which now start to join, with increased
frequency, with hydrogen and each other. Hydrogen and carbon have adaptable landing docks,
and together with oxygen and, to a lesser degree
also with nitrogen, calcium, phosphorus and
several other atoms, can form large and giant
molecules that will one day enable *life*. That's why
they'll be called organic molecules. A simple
sugar molecule with 24 atoms of 3 elements
($C_6H_{12}O_6$), or a not very large molecule of one of
the fatty acids with 72 atoms of 3 elements
($C_{18}H_{34}O_{20}$) belongs among them. Smaller organic
molecules (monomers) have the ability to form
long chains of giant molecules (polymers) with
hundreds of atoms of 3, 4 or 5 elements.

The organic molecules form materials with a wide range of properties, colours, textures and physical and chemical characteristics, depending on the sequence and connections between individual atoms. Just a few elements enable the formation of a huge range of extremely dissimilar materials. Some of the arising molecules have abilities that will once prove themselves outright miraculous. Among them, for instance, is adenine – a molecule with 11 atoms of 3 elements: $C_5H_2N_4$.

*BAY OF LIFE.* Somewhere at the shore of a primeval sea, the movements of the Earth's plates and the erosive action of water pouring from primeval mountains have already been forming a picturesque cove for millions of years. It may be a very big cove, or it may be quite small, but most likely it is shallow and almost completely separated from the primeval ocean by rocky headlands. A crack in the Earth's crust is stretching across the bottom of the cove, and it emits hot gases (at times) from red-hot magma of Earth's depth into warm waters of the cove. Some of the gases bubble through into the atmosphere; others dissolve into the warm water, into a saturated chemical soup of the bay of life. It is already about 10.6 billion years after the big bang and, by now, time is nearly ripe for the biggest feat of Nature.

The water of the cove is saturated by small molecules with atoms of carbon, hydrogen and oxygen, and occasionally other atoms. Within the sea currents, these molecules often collide, and now and then are joined together to form new and bigger molecules. Very slowly the number of unique molecules is growing. One of them has got a foaming quality. It forms a very thin membrane. Other molecules – under the influence of gentle wave action on a sandy beach – inflate this membrane into microscopic bubbles of foam. For some reason, molecules penetrate into the bubbles easily, but it's more difficult for them to get out again. That is why a diversity of molecules inside the bubbles is increasing. Some molecules unite; then they split again. The water foams; the bubbles change; they become larger or smaller, afterwards they burst again. Nothing much else happens for millions of years, but then suddenly …

MATH RULES. Sun rays of a new day spill over the glittering surface of a puddle, which has frozen overnight into a hard ice. The rays provide warmth, the ice softens, but it remains ice. Then quite suddenly, within one degree of Centigrade, the ice dissolves into cool water. Somewhere else water of a puddle covers a deep crack in the Earth's surface. Hot gases sputter from the crack and warm-up the puddle. Swiftly, water fizzles and evaporates into invisible vapour. Water can go from 0°C to nearly 100°C while only a little bit of water evaporates. Why is it then that water completely vanishes from liquid to vapour when it gets to 100°C?

Mathematics helps to explain many natural phenomena. The correlation known as random graph lurks behind changes of states of matter, as well as – just to use a simpler example – behind grouping buttons with threads: First, scatter 20 buttons on a table. Then, quite randomly and blindly, start tying the buttons with pieces of thread. After each tying action, establish the size of the largest group of buttons and record it on the vertical axes of the graph. On the horizontal axes of the graph, record the number of threads used so far.

The first piece of thread ties 2 buttons. Therefore on the vertical axes is 2, on the horizontal axes is 1. Because tying of the buttons is entirely random, it is highly probable that the next

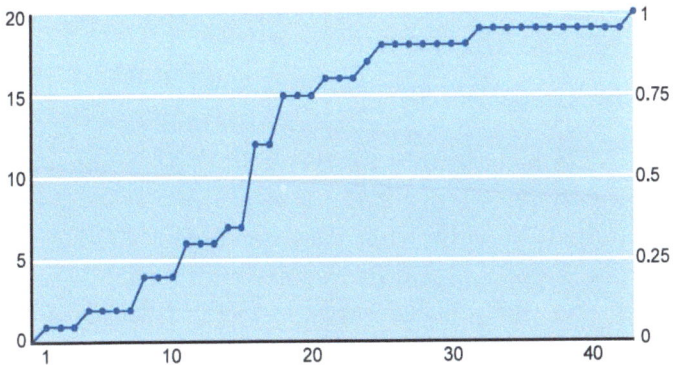

Pic.1

2 connected buttons are not connected with any of the first 2 connected buttons. The largest group, therefore, still has 2 buttons, but the number of threads has increased to 2.

When all 20 buttons are tied together, the graph looks more or less like picture 1. If we tie the buttons again in the same random way, the graph would look similar, just a bit different.

In order to get a graph easily comparable to other graphs, we can successively record – on the vertical axes – the number of buttons in the largest group divided by the total number of buttons. The largest number thus obtained will be 20/20=1.

If we tried to illustrate how much water in a puddle will transform into vapour at different temperatures, we get something like picture 2.

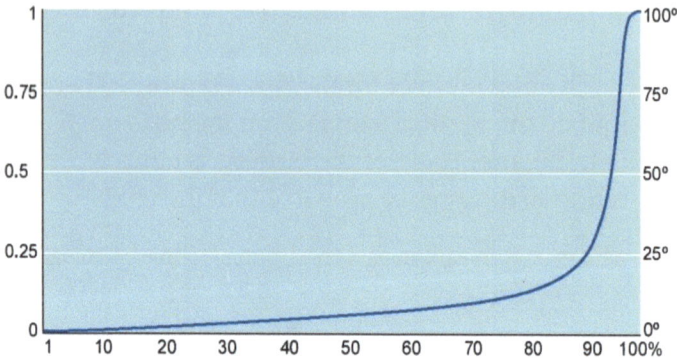

Pic.2

All of these graphs have one thing in common. In a very narrow zone, the measured value grows extremely steeply. This phenomenon is called the phase transition, and the results are not surprising. The phase transition is predictably reflected in any random graph.[2]

Changes in the state of matter are examples of the phase transition. But now we shall return to the foaming water of the bay of life.

CRYSTALLIZATION OF LIFE. Some pairs of organic molecules can't join, even if they collide in an environment well suited for chemical reactions. Then, however, a third molecule appears – a match-

---

[2]  Stuart A. Kauffman; *At Home in the Universe*

maker of sorts – and holds one of the molecules in such a way that the previously unsuitable pair easily joins. The matchmaker stays free to help in the joining of other pairs. The matchmaker will be called a catalyst.

For millions of years, the number of unique molecules is increasing. The number of molecules with catalytic ability is increasing as well. That's why, quite suddenly, the array of molecules in one of the bubbles reaches a phase transition, and the number of molecules shoots up considerably. The bubble doesn't contain only a few thousand unique molecules anymore; it now hosts more than 100,000 types of molecules. The primeval soup of life is becoming a complicated self-catalyzing chemical concoction[3], which will probably never become fully understandable.

But the soup of life is not alive yet. It will take another few million years before some molecules will coincidently gather in a special way, and something new and extraordinary happens. One of the bubbles prolongs itself and divides into two exactly identical bubbles that, in a self-controlled process, will soon go through the same process again. These bubbles – primeval cells of a sort – all of a sudden live and absorb into themselves the ambient chemical soup as food and periodically reproduce themselves. More importantly, they also multiply and, therefore, soon outnumber the lifeless bubbles, which can only inflate and disappear again by bursting into insignificance.

---

[3]  Stuart A. Kauffman; *At Home in the Universe*

Perhaps only Nature knows exactly how life began and how it works. It is too complicated. In spite of that, with the help of appropriate simplifications, it is possible, even without much of proper education, to get a glimpse into the complicated workings of life.

PLAN OF LIFE. One of the basic prerequisites to a successful life of primeval cells, and all future life forms, is the ability of unfailing, always identical self-reproduction. The formation of nucleic bases (nucleotides) made such a self-recreation possible. Four fundamental nucleic bases (adenine, cytosine, guanine and thymine) easily join with the help of less important building blocks derived from molecules of sugar and

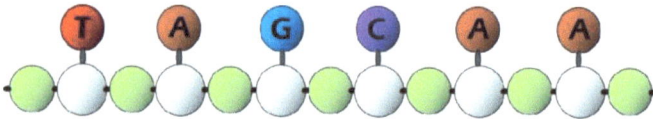

Pic.3

phosphoric acid. Together they form long chains in any sequence (Pic.3) that will be called DNA. These nucleic bases can also tie themselves together in yet another way: with a weak bond of 2 or 3 atoms of hydrogen. But the hydrogen-bridge can only tie adenine with thymine, and cytosine with guanine; no other base pairs are possible.

Thanks to this coupling ability, nucleic bases form chains that combine into the double helix DNA – sort of a ladder. One of the spirals of the helix actually creates a sophisticated language of Nature. It will be called the parent strand. In pic.4, Nature says "tag caa" in the parent strand. The second opposite chain isn't a part of the language, but it belongs to a copying and communication mechanism, and it will be called the matrix strand. In reality, the chains are not

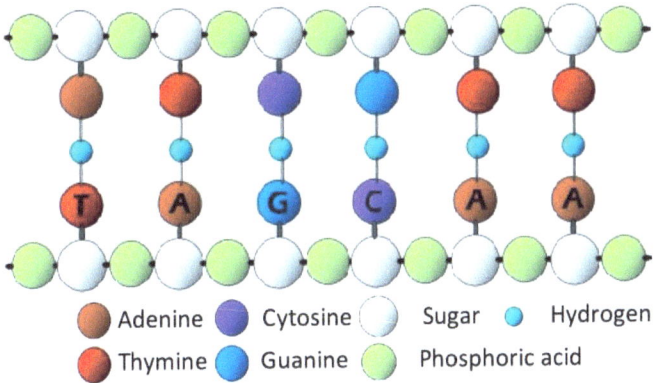

| Adenine | Cytosine | Sugar | Hydrogen |
| Thymine | Guanine | Phosphoric acid |

Pic.4

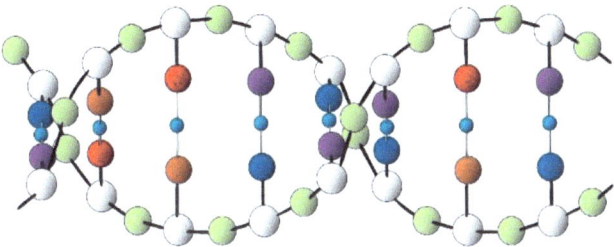

Pic.5

straight but are firmly twisted (Pic.5). The whole DNA is not in one continuous helix; it is divided into several sections called chromosomes.[4]

In order to fit a lot of information into each chromosome, winding of the helix is extremely tight. Many long sentences fit into each primeval

---

4  H. Reinbothe; *Molekül, Mikrobe, Mensch*

cell. All secrets of life are recorded in those sentences. Each primeval cell carries inside the double, spiral-shaped plan of life.

The primeval cells float in the water of the bay of life and ingest molecular nutrients through their cell membrane. Inside, the nutrients are soon chemically transformed into a building material for the future primeval cells. When there is enough of the building material, the weak hydrogen bridges break up, and the double helix separates into two single spirals. The primeval cell now divides itself into two cells in such a way that each cell contains one single DNA spiral. Thanks to the limited coupling ability of the nucleic bases, bits of the building material now connect with the single spiral in each new cell until it is doubled up, exactly as it was in the mother cell. That's the way, approximately, that the copying system of Nature works. That's how the plan of life gets passed on to each new cell.

DIRECTING ORDERS. Another basic requirement for a successful life is the ability of the DNA spiral to reliably direct and influence all enormously numerous life processes. This is enabled primarily by a molecular messenger, which is known as RNA, and is derived from DNA. The matrix strand of the DNA double helix becomes temporarily disconnected at one point. The nucleic bases attached to the matrix strand in this area thus become exposed to the molecular building material available inside the cell. The suitable nucleic bases (for example CAA (Pic.6)) now connect to the matrix strand to form an RNA messenger. The messenger is now finished; it tears off the matrix strand and floats to the location inside the cell where proteins are built. At the building site,

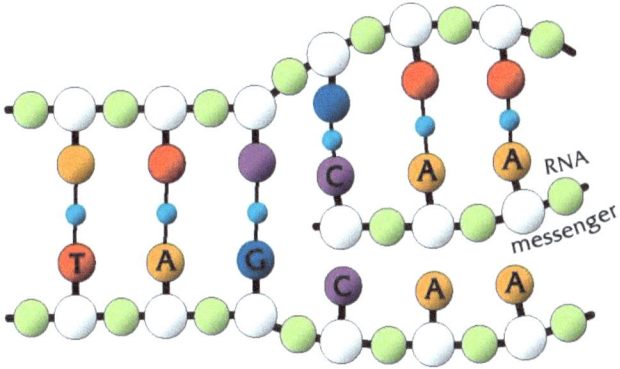

Pic.6

CAA is a command to build the amino acid called glutamine. The group of molecules CAA is a sort of catalyst that can catalyze only glutamine. The next moment another messenger floats in, say CCC. CCC can catalyze only alanine; so, it is manufactured and connected to glutamine. Messenger after messenger directs an exact sequence of amino acids in a long, twisted and often branched-up chain that makes up a giant molecule of a protein.

DNA language has only 4 letters (A, C, G, T), and forms only 3 letter words. When only 4 letters are available, it's possible to make 64 three letter words. There are only 20 amino acids; the DNA language is, therefore, more than sufficient for making all of them. Some words are even wasted as a duplicate command for making one amino acid. Other words serve as a command to end the manufacturing of one protein and to start another.

Three atoms $SiO_2$ form a quartz molecule, a major component of granite rock. Three atoms $H_2O$ are water. That's a small sample of a variety and diversity offered by different combinations of only three atoms out of about 110 found in nature. 20 amino acids arranged in different sequences into giant molecules of proteins with hundreds of atoms offer a variety and diversity that is utterly incredible.

Probably only a few hundred proteins are in the primeval cell. The human body will contain at least 200,000 proteins, perhaps even several million, out of which only about 25,000 proteins are known today. Eventually, in the entire biosphere, there will be a trillion proteins.

STILLNESS, CHAOS, ORDER. Several large molecules emerged from the soup of life at one or more spots in the bay of life. These molecules do not integrate well with other molecules; they are also quite resistant to disintegration. They form a substance that restfully descends to the bottom, where it remains in rigid stillness.

But energetic chemical action persists at various locations of the bay. About 100,000 different kinds of molecules have reacted to each other for hundreds of million of years already. Chemical matchmakers are busy helping others join. Some molecules are getting bigger, and then they shatter again in random collisions. Chemically speaking, a confused bustle, utter chaos reigns here.

Only at one spot of the bay, inside the primeval cell, did the chemical chaos jump into magical order. The primordial cell absorbs chaotic ambient molecules through its membrane. Inside the cell, helix DNA forms RNA messengers that float to the cellular construction site and gradually give rise to complex protein molecules. The primeval cell grows and divides into two cells, then the whole process repeats itself – in perfect order – again and again. The miraculous order of life reigns here.

How miraculous is this order? What probability was there that the chemical chaos would jump into the perfect order? Infinitesimal? Really high? Was it inevitable? Nobody knows. But one thing is certain; lots of time was required. Given so much time, the impossible became possible, the possible probable, and the probable virtually certain. [5]

There is an extremely complicated self-catalyzing net of organic molecules in the primeval soup of life. This complexity was certainly instrumental for the origin of life, but it doesn't make it easy for us to see if the rules of Nature must lead to the origin of life. The work of professor Kauffman indicates that it must. (To learn a bit more about this work, please go to page 98, Appendix 1.)

In large chaotic systems, there are many components that can keep changing. The sum of these changes doesn't lead to an overall change of the system most of the time, but the components keep changing. Provided they keep changing for a sufficiently long time, the sum of the changes can

---

[5] George Wald; *The Physics and Chemistry of Life*

kick-start a spontaneous crystallization of mirac-
ulously improbable order.  Favorable conditions
for the genesis of the right chemical soup of life do
not exist everywhere.  But where they do, the
genesis of life might be absolutely inevitable.[6] It
just takes a long time.  For us it is an immensely
long time, but in the cosmic scale, it's not all that
long; this is nicely expressed by this joke:

A child talks to God.  "How long is million
years for you?"
"Just as one second is for you."
"And how much is million dollars for you?"
"Just as one cent is for you."
"Only one cent!  Could you then give me one of
your cents?"
"Yes I could.  Just wait a second."

[6] Stuart A. Kauffman; *At Home in the Universe*

# EVOLUTION

*NOTHING IS INFALLIBLE.* Primeval cells are replicating. Each new cell knows exactly what to do because it has DNA helix that, with the help of RNA messengers, directs all the processes of life. Three letter words of the DNA chain are divided by starting and finishing words into individual segments that direct specific processes. These segments will be called genes. The primeval cell does not have more than a thousand genes, perhaps only a few hundred.

Replication of DNA is extremely reliable. Many millions of primeval cells were already born, and the helix is still exactly the same. Until, all of a sudden, one word in one gene accidentally changes, due to a replication mistake. This kind of mistake will be known as mutation. Perhaps the first mutation affected the construction of a protein for the cell membrane. The membrane is now a bit stronger than before, but, unfortunately, food can only get through it very slowly. The primeval cell with the new mutation grows slower, and multiplies less often. The original faster growing cells vastly outnumber the new cells with the mutation, until the new cells die off altogether, and the unsuccessful mutation disappears with them.

Thousands of years go by, and one genetic word mutates again by mistake in one primeval cell. Again, it is a word that influences construction of the cell membrane. But this time the new mutation improves the reception of food. The advantageously influenced cell prospers well, matures quickly and produces a strong offspring that easily takes away food from adjacent cells. The lucky mutation spreads steadily throughout the whole population until all individuals without the successful mutation are pushed out of the population. That's how natural selection works.

Just as Nature didn't have to think up new atoms and molecules when they were first formed, she now doesn't have to plan for new mutations. That would be too much work. Mutations spring up accidentally without help. When they are well suited for their environment, they contribute to an easier life for their carriers, and they survive; we'll call them adaptive changes or mutations. If they do not fit to the environment of their carriers, we'll call them non-adaptive mutations.

Not all mistakes are mistakes. Some mutational mistakes lead to gradual improvements of processes and forms of life. Other mistakes cause quite revolutionary leaps for the better. Some mutations are neutral; they do not lead to improvement or deterioration. These mutations survive, and they can become deadly or useful later, due to substantial changes of the environment where hosts of these mutations live.

Thousands of primeval cells are transformed and killed off due to unsuccessful mutations. In spite of that, primeval cells grow in numbers, and one day there is not enough food for all of them. Then, in one cell, a new gene (several genetic words) is accidentally added to the GNA helix. The new gene leads to the construction of a new protein that grows a little tail at one end of the cell. This new protein is quite restless and it undulates constantly. The movement of the tail propels the cell through the soup of life. Actually it's not a primeval cell any longer, but a kind of slipper animalcule. The mobile slipper bumps into food molecules more often, and thus wins over primeval cells in the constant chase for food.

Not all proteins catalyzed by RNA messengers are used for the body-building of live organisms. Some of them assist in hundreds of chemical conversions in the digestion of food and other processes of life. Some of the special catalyzing proteins will be called enzymes.

Another mutation now leads to the creation of another kind of protein – a hormone – that will be used in the influencing of the behavior of other proteins. Hormones are chemical messengers that travel from the place where they were constructed to a place where their influence is required.[7]

Only one slipper is now endowed by the gene for the making of the first hormone. This hormone is only made when the slipper had enough to eat and is pleasantly full. The hormone quickly spreads throughout the small slipper's body all the way to the slipper's tail. Under the influence of the hormone, the tail relaxes and stops moving. The satisfied, pliable slipper is now carried only by ocean currents. The slipper doesn't waste energy required for swimming while it doesn't need to absorb any food for a while. The ability to save

---

[7]  H. Reinbothe; *Molekül, Mikrobe, Mensch*

energy is very beneficial, so it quickly spreads through the slipper population.

Genes have only three basic functions or abilities: most of the time, they reproduce themselves correctly. Secondly, they make RNA messengers, and through them, they direct construction of all living matter, including enzymes and hormones, which help genes to direct the behavior of living organisms. When they copy themselves, they work extremely precisely. But exceedingly rarely, errors in DNA replication do happen. These accidental errors – together with natural selection – ensure the evolution of nature. Making errors in its own replication is the third function of genes.

Millions of years pass by, and hundreds of millions of unsuccessful mutations kill off their unlucky hosts. Yet, hundreds of successful mutations lead to an incredible variety of organic inventions. One of them is the origin of multi-cell organisms. Now, life processes don't have to be taken care of by only one cell. Specialized cells receive nutrients and send them to other cells which, for example, undulate and provide propulsion for the whole organism. Still others form specialized organs of continuously improving creatures.

The population of organisms grows until there is not enough food for them all. The genetic mutations and the food scarcity in the primordial soup of life lead to the origin of plants and herbs. They take root at the bottom of the bay and draw nutrients from a growing layer of dead and decaying animate beings. Favourable mutations lead to the beginning of herbivorous animals that can eat plants. Another set of mutations leads to the origin of predators who catch and devour herbivores. Life is becoming a complicated ecosystem with many mutually dependent species of animals and plants. The cells of all these species contain exactly identical molecules of DNA and RNA. That's why it seems possible that absolutely all organisms – that already exist on Earth and that will ever exist – have their origin in the same first primeval cell of life. But it may not be necessarily so.

Metabolism is a complicated chemical process within the cells that maintain life and release energy from nutrients. The energy is used for the construction of proteins, movement, and all other life processes. When an organism's metabolism is improved, the organism requires less food to maintain life. Many mutations now lead to the creation of a specialized digestion system that improves the efficiency of metabolism.

Complicated organs develop gradually over millions of years. For example, light-sensitive cells appear by chance at the front end of a grub or a caterpillar. Thus endowed grub can now tell when a body of her enemy casts a shadow on her. That provides a signal for her to start moving fast. With a bit of luck, she escapes the blind predator. The next mutation creates a sharply defined recess that contains the light-sensitive cells. The edge of the recess provides protection for the important cells, but the shadow cast by the edge will prove even more important than protection. The grub can now feel where the light is coming from. Gradually, the recess deepens to form a spherical cavity with only a small round hole above it. Light coming through the hole paints a vague picture on the sensitive cells. Now it's only a matter of a few hundred thousand years before the hole above the eye cavity is filled in by a lens made of ingenious transparent protein that sharpens the picture.[8] Each evolutionary step must bring some advantage to the owner of the new developing feature in order for the feature to be retained by natural selection.

All of these changes are accompanied by the gradual development of nerves and the central nervous system that can produce and transmit weak electric pulses. Electric signals greatly complement the signaling ability of hormones. The nervous system is now able to transmit, receive and interpret the meaning of the signals from the light-sensitive cells in the eye and signals from other developing sensing organs.

Other animal body features develop abilities that seemingly contradict some rules of Nature. In spite of gravitation, a lizard gecko easily walks on smooth vertical surfaces. The gecko does not have sharp claws capable of hooking into hard surfaces, nor special glue on its paw pads. For the gecko, evolution developed something far more ingenious. The soles of its feet are intricately shaped and covered with forward facing hairs so fine that about two billion of them fit into one square centimeter. When these hairs are pushed in one direction, they penetrate in between molecules of the

---

[8] Richard Dawkins; *The Blind Watchmaker*

surface they touch, where they are within the influence of van der Waals intermolecular force. This force is then capable of holding the gecko on perfectly smooth vertical, even overhanging surfaces. Pulled in another direction in the next step, the hair of the gecko's feet escape the influence of van der Waals force once again. Only Nature knows why these extra fine hairs do not break or get hopelessly clogged up by dirt.

   Ordinary mistakes in the replication of genes, random additions of new genes to already time-tested genes, together with natural selection, and a very, very long time constitute a coughing engine of biological evolution that tirelessly created everything from the primeval cell to the multitude of organisms with incredibly complex and mostly flawless bodily organs, metabolic processes and interesting ways of life. The principal of evolution is remarkably simple and highly effective. Even so, it does not lead in a direct line from simple and primitive to complicated and sophisticated. It travels a rough and long way through the fitness landscape.[9]

FITNESS LANDSCAPE.   The ability of an organism to cope with its environment can be called its evolutionary fitness. The elevation and location of each organism in the fitness landscape represents its current level of fitness.

   There isn't just one fitness landscape. Each species of organisms has got its own fitness

---

[9]  Sewall Green Wright; Evolution & the Genetics of Populations

landscape. An organism moves through its fitness landscape depending on what mutation – a chemical change in its DNA helix – just happened. The higher fitness an organism achieves, the higher it climbs in the landscape. The total number of mutations any species can go through is given by the size of the landscape.

The size of the landscape is given by the number of genes an organism has, and number of states (alleles) each gene controls. For example, one gene can make something green or brown, another can make something smooth or rough, and so on. A gene can have two or more alleles. If an organism has only 3 genes, each with 2 alleles, then this organism could mutate to only 8 variations. The fitness landscape of this theoretical (and impossible) species would have only 8 points, say 5 hills and 3 valleys.

Each genotype can mutate in one step only to 1 out of 2 or 3 potential genotypes, kind of chemical neighbours. That means that a mutation can move a genotype only to a few neighbouring points of the fitness landscape; it can't catapult a genotype to a far away point.

Unlike in the real Earth landscape, in the fitness landscape it is easier to go up than down. The unlucky few who go down to lower fitness are usually pushed out of the population by natural selection. That makes crossing to the other side of a low valley, where the higher mountains might be located, difficult or impossible.

If you'd like to see more detailed information about the concept of the fitness landscape, please go to page 103, Appendix 2.

Mutations sometimes enlarge the number of genes. Each new gene doubles the size of the fitness landscape. For a species with four genes, each with two alleles, the path of evolutionary fitness would lead to 16 points with different elevations. Five genes with two alleles would enable 32 genotypes. 2, 4, 8, 16, 32 – that's the beginning of exponential progression. It does not seem that it grows too quickly.

For an illustration of how the exponential progression grows, we'll turn to an old legend about the inventor of the game of chess, who taught his new game to his monarch. The monarch liked chess very much and suggested to the inventor that he choose his own reward for the game. The inventor chose wheat: one grain at the first square of the chessboard, two grains at the second, always a double amount of grain at each subsequent square. The monarch praised the inventor for his modesty and called his mathematician to calculate the whole amount. The calculation took a surprisingly long time to make, and then the monarch was utterly astonished by the result: there was not enough wheat in the whole country to pay the named price. There are 64 squares on the chessboard. On the last square, there should be nearly 18.5 quintillion grains of wheat; the exact number actually is 18,446,744,073,709,551,616.

The magnitude of this number is not easy for the human mind to comprehend. We can easily relate to things that surround us daily. We can approximately imagine, say, a 3,000 metre tall mountain, even though we don't see it daily. But we already saw it on a picture at least. But how tall would be, say, a 100 billion millimetre tall mountain? 100,000,000,000mm = 100,000,000m. A picture of such a mountain doesn't appear in our mind's eye.

Yet, 100 billion is a ridiculously small number compared to 18.5 quintillions, which is a potential number of genotypes for an organism with 64 genes, each with 2 states (alleles). The fitness landscape of this organism would have hundreds of millions of hills, mountains and valleys. But a clump of proteins formed by 64 genes is ridiculously small. It is not sufficiently large for crystallization of life. The smallest known organism (pleuromona) has several hundred, maybe a thousand genes. Let's say that the smallest possible number of genes in a living organism is 500. Some genes might have more than 2 alleles, but let's say that all of the genes have only got two alleles. 500 genes enable such an enormous number of genotypes that it is pointless trying to imagine this number.

Now let's imagine the fitness landscape of Homo sapiens, which has about 100,000 genes. This fitness landscape has an unimaginable number of peaks and valleys that stretch in front of an evolutionary tourist to a nearly infinite distance. Each mountain, even a small hill, is a trap that prevents the evolution of organisms to higher fitness and more sophisticated perfection.

So, how is it possible that there was evolution
from the very primitive to the extremely sophisticated? One of the answers to this puzzle is sexual
reproduction.

*SEX.* Offspring, or progeny of asexual reproduction arise from a single parent and inherit genes
from that parent only. It is genetically identical to
its parent, unless it is a mutant due to a mistaken
transfer of genes. That's why there is a low genetic diversity in any population of asexual organisms. Almost the entire population has the same
genotype – it is found in the same location of the
fitness landscape. Only mutants walk over to the
neighbouring points.

Sexually reproducing plants and animals contain two sets of genes in all of their cells: one set
comes from the mother and the other one from the
father. Only in reproductive cells – in eggs and
sperms or pollen – is there a single set of genes
created by a random combination of genes from
its parents.

At fertilization, the single sets of genes of the
parents' two reproductive cells join up in the first
cell (zygote) of the new offspring. Picture 7 sche-
matically shows how genes are mixed up when
reproductive cells are formed. In real life, the gene
combining process is more complicated than that,
since genes from a grandmother and grandfather,
carried by the parents' DNA, are not all at the
same place as the picture shows, but they are at
different places of the DNA helix that is divided
into as many as 26 chromosomes. Perhaps only

Nature knows exactly how the genes are mixed
without a mistake and transferred to each new
generation.    The mixing of genes produces tre-
mendous genetic variation in any population of
sexually

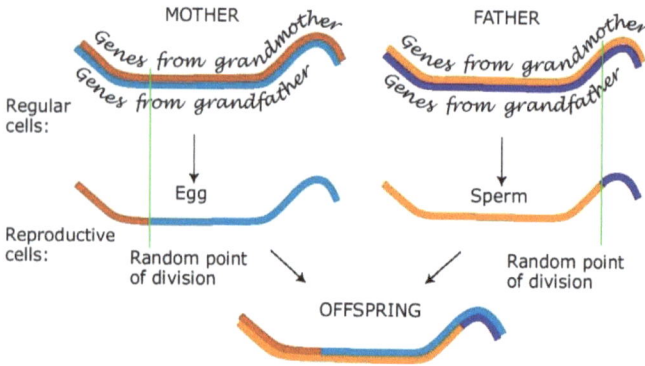

Pic.7

reproducing organisms.  In the fitness landscape,
both parents and their offspring are at different
landscape points.  Only identical twins are in the
same point of the landscape, because they receive
exactly the same copy of their parents' genes.

Mutations occur extremely seldom, and they
move the progeny from the parents' location to
adjacent points of the landscape.  But absolutely
every product of sexual reproduction is moved to
adjacent points or even catapulted to more distant
landscape locations.  That's why sexual reproduction greatly accelerates the speed of evolution.
Offspring usually have similar fitness as their parents, thus they are at approximately the same elevation in the fitness landscape somewhere in sight
of their parents.  But the parents may be in a location where no path leads to higher fitness.  They

are trapped at summits of small hills where all trails lead to lower fitness. However, with the help of genetic mixing, their descendants may be catapulted over a valley to a location from which several mutation paths lead to a considerably higher fitness level.

For an asexual organism, a mutational trap on a low summit means practically the end of the entire species. Such trapped species can't evolve any longer, and it is only a matter of time before they are squeezed out from the evolutionary scene by other more successful species. However, for sexual organisms, a stay on a low hill of the fitness landscape is only an opportunity for sightseeing, a short rest and a pleasant occasion for sexual reproduction. Sexual procreation moves evolutionary development much faster than infrequent mutations. No wonder most organisms procreate by sexual ways.

MATH STILL RULES. Sexual reproduction is not the only accelerator of evolution. Mysterious relations of math with live organisms also play a big role in the speed of species evolution.

Let's imagine, for example, an advanced animal with tens of thousands of genes, in other words, with the fitness landscape of hundreds of millions of peaks and valleys. By chance, these peaks are uniformly spread throughout the landscape. Low and high hills are evenly mixed with high rolling plains and flat lowlands. Surprisingly, this animal will not develop too far despite the fact that it reproduces sexually. Evolution could get this ani-

mal from a low point to the highest peaks of this
landscape, but it can be calculated[10] that it would
take much longer than the entire duration of the
universe so far.  Due to its slow development, this
animal will be outpaced by many fast improving
creatures.

But what happens if the landscape's hills and
mountains are not spread out uniformly?  What if
the fitness landscape is arranged – quite accidentally – the same way as the real Earth landscape? Here, low hills are concentrated into a
small, gently rolling highland behind a spacious
lowland; elsewhere, jagged, high mountains tower
over a flat but high plateau.  Skillful mathemati-
cians can calculate again, how many random steps
evolution must take to get from the low trailhead
all the way to the landscape's highest peaks, and
how many generations it will take.  Purely due to
mathematical reasons, evolution will run through
this landscape very quickly.  Accidentally, some
species have better fitness landscape than others.
Many mammals, including humans, will have a
lucky fitness landscape; that's why men will
evolve from the first mammal in a relatively short
time of about 100 million years.

MANY KEY TURNING POINTS.  The first organisms do not need oxygen for their metabolism.
That's entirely predictable since there is very little
available oxygen in the atmosphere or water.  Almost all oxygen is tied up with various other at-

---

[10] Stuart A. Kauffmann; *At Home in the Universe*

oms in various molecules. Metabolism of the first
organisms breaks up molecules that contain oxygen that is then released as a residual product.
This is how pure oxygen slowly accumulates in
water and atmosphere. Then, over time, pure oxygen enables the genesis of an entirely new metabolic system that uses oxygen as an important re-
source and releases carbon dioxide as a residual
waste. The oxygen-based metabolism is quickly
proven as a key turning point in the evolution of
life since it is very effective, and it compliments
the metabolism of plants (the photosynthesis).
The photosynthesis needs carbon dioxide as a re-
source and releases oxygen as a waste.

Water is the ideal environment for life. Among
other things, it easily eliminates climatic extremes.
Even the scorching sun does not heat up the
depths of oceans and lakes. Occasional night frost
hardly changes the temperature of lakes and seas.
Far away from the equator, where dry land will
remain barren land void of any life for hundreds
of millions of years, bitingly deep freeze does not
influence vigorous life forms proliferating under
thick ice. Yet, life soon forces its way from water
to shore. For super small bacteria, it's easy to hide
even on dry land and escape from heat or frost, so
they are perhaps the first to crawl onto dry land.
Life is spreading and branching, but very slowly.
Nothing much happens until the appearance of
sexual reproduction (the already mentioned accelerator), which enables faster and more varied development about two billion years BP (before present).

So the remote ancestor of molds, moss, and dry land plants says goodbye to the water environment about a billion or perhaps 800 million years BP. About 300 million years later, the ancestors of mollusks and insects find richer sources of food on dry land. Vertebrates also originated in the oceans, but they too forced their way to dryer environments about 400 million years BP, but some of them return to the oceans millions of years later.

Rooted plants can't run away from the scorching sun or frost, so they stay for a long time only in moderately warm regions, often in the shadows of rocky outcrops. Eventually, green stalks become woody due to a mutation, and other mutations cause improvements that enable the expansion of plants in very cold regions.

Small mobile creatures are able to hide from temperature extremes and unfavourable weather; among them are insects and amphibians endowed with an ingenious, evolutionarily random invention. Their life cycle consists of several entirely different forms. For instance, nymphs of a dragonfly are born from fertilized eggs under water, where they are protected from temperature extremes. There, they voraciously devour various underwater creatures for nearly a year. Other kinds of dragonflies live under water as long as four years. Then, on a warm summer day, a dragonfly nymph somehow feels that her underwater hunting life has neared an end. Driven by unstoppable instinct, she climbs onto a sun-warmed rock on the shore. Her skin warms-up pleasantly, dries-up and finally cracks. The former nymph, covered with a brand new skin, climbs out of the crack. On her back are four small bumps, folded wings interlaced with fine veins. The veins are now flooded under pressure with a clear fluid and the bumps slowly unfold into transparent, flat, prettily-coloured wings. In about twenty minutes, the new dragonfly can fly about in a summer breeze, as the transformation from an underwater hunter into a predatory flying acrobat is completed.

Not all dragonflies survive this transformation. Hundreds of them are devoured by other predators while their wings are helplessly unfinished; but thousand now make a living by hunting on their wings. They inherit 3 instinctive life goals from their ancestors: they must try to avoid predators that could devour them, they must hunt and eat to gain strength and complete their growth, and they must find a partner for mating. Only Nature knows how they find a mating partner. Perhaps finding a partner is totally random, or they recognize one which is sexually compatible with them. Maybe they identify those who fly the best. One way or another, a female now chooses a male. At the back end of her long abdomen, she has pincers, and uses them to grab the male right behind his head. Now they fly together. She is in the front, he, involuntarily, behind her. Finally he gets it: she won't let him go until he fertilizes her.

Before the onset of colder weather, she lays her fertilized eggs onto a water plant or directly into

the water before dying.  The life cycle ends and
begins again.

Sensory organs continue developing and, coordinated by the central nervous system, they are
used to find food and partners for mating, and
help to escape the reach of predators.  Other mutations lead to an ability of males to block access to
females by other males.  So, for example, one beetle gains an ability to seal up the sexual opening of
a female after mating.  Females of this beetle usually mate several times, but the sealed up female is
not able to successfully mate with another partner.
The sealing ability has proven itself an adaptable
change, so it spreads quickly throughout the
population of this beetle.

Males of other species have to fight for females
in ritual or real, often brutal, fights.  In this way
only the strongest males pass their genes to future
generations.  Fighting for females is a better evolutionary invention than the sealing up of females.

The most important development for utilization
of regions with significant temperature extremes is
perhaps the origin of warm-blooded metabolism.
To maintain their body temperature at a level necessary for maintaining of life, warm-blooded animals don't have to rely on the environment.  From
their food, they can make their own heat when it's
cold.  When it's hot, blood circulating in the skin is
an efficient heat exchanger for the cooling of inner
organs.

Another important evolutionary invention is
the territorial instinct.  The strongest individuals
seize up good areas for themselves.  Subcon-

sciously, they mark it mostly by scent marks, or by visible marks, such as clawed up bark. Although the owner of a good territory – mostly a strong male – must fight for a territory with whoever is courageous enough, he doesn't have to fight for food and often not even for females. Females recognize the advantages of good territories, and they voluntarily join the lucky owners. To avoid strong landlords, weak males remain at less valuable fringes of good territories. Here, life is not so easy, but they are safe, and they can substitute for a suddenly perished landlord if they are lucky. Many songbird males clearly fight for territories, not for females, and mark their territories by incessant singing.[11]

Then, at some point, memory comes into existence. Now, for the first time in the history of life on Earth, a living organism endowed by memory can react to its environment not only instinctively – guided by instructions encoded in genes. Experiences stored in memory now become an important guideline in the struggle for survival.

Memory has a role in many inventions, for instance in the ability to choose more variable food. The metabolism of each species is capable of using only a limited number of molecules from trillions of proteins and ten million organic molecules that exist in the biosphere. An ability to recognize molecules suitable for food must have been encoded in genes for a long time because consuming unsuitable molecules leads to a waste of energy at

---

[11] Robert Ardrey; *African Genesis*

best, or to a quick death by poisoning at worst. Genetically-based identification of suitable food usually leads to utilizing only one kind of food. When it comes to the choice of food, primitive organisms are specialists. Memory, in combination with the trait of curiosity, leads to the very first omnivorous behaviour.

So, for instance a prehistoric bear male – carnivorous specialist – under pressure from the scarcity of his usual prey and guided by curiosity, tries to eat an interesting smelling plant. This experiment keeps him alive, and his memory helps him to find this new kind of food in the future. Unfortunately for his bear cubs, this experience is good for nothing. In contrast to genetic information, the information in his memory is not transferred to his offspring, and the male bear is not involved in bear cubs' upbringing. But new experiences collected by a female bear's memory are transferred to her bear cubs because they remain under the care and supervision of their mother for a year and easily learn everything she knows. As millions of years go by, the quantity of information in a bears' memory gets larger. One year is not long enough anymore for teaching bear cubs everything they need to know, so they stay with her mother gradually longer, eventually two, even three years.

Further development of the memory and central nervous system enables close cooperation of individuals within a species whether in foraging or in defensive actions against natural enemies. Communal living has its origin already in some insects. Cooperation among ants or bees is admirably complicated, yet it is entirely automatic, encoded right in their genes, is inborn, reflexive and unchanging. It remains exactly the same, from birth to death, from one generation to the next, for hundreds of millions of years.

Now, cooperative way of living is both instinctive and memory based. Solitary ancestors have noticed the advantages stemming from cooperation during short mating periods, and stay together even after mating. The origin of cooperation is an adaptive change for many species because membership in a group brings advantages to individuals. If it doesn't, an accidentally formed group falls apart again, and their social way of living doesn't develop.

So, bears, squirrels or tigers remain alone and independent unless they are mating. Wolfs, antelopes, lions or monkeys live in packs, herds, prides or troops. In contrast to ants and bees, the social life of some higher order of animals is not governed only by genes, but also by the transfer of experience and cooperative habits from one generation to the next. The genes determine biological design and metabolic processes, as well as the social cooperation of lower order animals. Experiences and habits that influence social life of higher order animals could perhaps be called "sogenes". Then, the changes of old habits and the origin of new habits and experiences could be called "sogenetic" mutations.

Genes are automatically transferred to future generations. For a gene to disappear, the whole population that carries that gene must die-off. However, each generation must learn sogenes again and again. When one generation of a group stops using a sogen, it is immediately lost for all future generations of this group.

Transfer of all memory-based skills to young becomes easier and faster within a social group. Compared to cubs of solitary bears, wolf or lion cubs gather experience not only from their mother but also from their father, siblings and other individuals that belong to their group. More or less the whole pack or pride is involved in the cubs' upbringing.

Communal living also fosters development of new skills. For instance, a recently formed pride of lions hunts its prey exactly the same way as its solitary ancestors. Then one day, a strong young lion chases after an antelope. Entirely by chance, they head towards a group of lionesses lazily resting under a tree. The noise of the chase wakes up the lionesses, and they treacherously ambush the antelope. The similar situation happens more and more frequently until this ambush hunting style becomes a regular habit of the pride. A mighty male lion chases its prey towards awaiting lionesses which, being smaller and more agile, easily catch the prey in a short sprint, and mercilessly end the hunt. The invention of ambush hunting is an adaptive change, so it is passed to the next generation. But changes like that do not happen very often. Lions, wolves, as well as monkeys, live in stable societies. Their life style is unchanged over many, many generations. Sogenetic mutations occur very seldom; however, when compared to the rate of change of genes, they occur extremely often.

But why do lions grow bigger than lionesses? Stronger individuals can appropriate for themselves more advantages by force than weak individuals. Squabbling for each advantage leads to fights and injuries, sometimes even death. That's not good for individuals, or for the group. But eventually, members of the group remember how strong each group member is, and they naturally fall into a hierarchical order. The strongest individual dominates over all others. Fights can then subside or cease altogether. The most dominant individual, that is the individual with the highest status, is entitled, without fights, to the best food, as well as the best resting place. If the strongest one is a male, he often also has the preferential access to all females for mating. For some species, this priority is so high that the dominant male is the only one that mates with all fertile females of the group, regardless whether they are small and weak or large and strong. That is why body size coupled with strength is an adaptive trait for males, but not for females. This gradually leads to a difference in size. Removal of weak males from the reproductive cycle also leads to a better quality offspring.

Females of most animal species are only fertile during a short period once a year. This convenient arrangement ensures timing of offspring birth into the most suitable period of each year. Females of social species generally come to heat (are fertile) one at a time, not all at once. That's why they don't have to fight for males. In order to ensure the best possible quality for their offspring, females present themselves just to the best male. Then he, alone, can be the father of the entire next generation of his group. This arrangement makes it difficult to recognize all the things that males fight for. Dominant stallions or elk bulls, which are surrounded by their loyal mares, or elk cows, undertake fierce fights with rival males. Quite obviously they fight for the status of being the strongest male, and losers are chased out from the best pasture. But it doesn't seem that they have to fight for females. Mares and cows voluntarily stay in the pasture of the strongest male.

For some species, other evolutionary tricks ensure good quality of offspring. Some females choose their mating partners according to visible traits. For instance, a peacock hen chooses a cock that displays impressive tail feathers. Quite understandably, a strong peacock – that is able to get the most food – develops the most extravagant tail feathers for attracting the strongest peahen. By opening a large and colourful tail fan, a peacock can also fend off other peacocks from fighting for desirable food. Clearly, a noticeable sign of fitness is a beneficial device for reducing the number of fights for various advantages and dominant position in a group.

A genetically-coded position of individuals in a group also exists in nature, but only for social insects. However, insects have spent a very, very long time waiting for a status forming mutation.

Higher-order animals, such as birds and mammals, start personally fighting for social status hundreds of millions of years after the origin of insects.

Another reduction of dangerous physical fights was assisted by sogenetic mutation leading to the fidelity of reproductive partners. In a flock of crows, adolescent youngsters rank themselves into a hierarchic order with the help of a few fights. Only then do they form reproductive pairs. Naturally, the most dominant male chooses the most dominant, available female. This union is for life. For the remainder of their lives, there is largely no need to fight for status, food or females. A pecking beak is a dangerous weapon; no wonder the formation of stable reproductive pairs and stable hierarchical position in the flock are adaptive changes. A superfluous female has the lowest status in her group. If the female of the dominant male dies, he is not allowed to lure away another dominant female from another male, but he can pair off with a single female that languishes discontentedly at the bottom of the flock hierarchy. Because a male partner is willing to fight for his partner's advantages, the formerly single female, now mated to the dominant male, becomes the most dominant female of the flock. She now doesn't have to fight for anything, but with memories of unsuccessful fights of her youth, she nevertheless looks for skirmishes. She is victorious now because everybody remembers and respects her new dominant position, and skirmishes then soon fade away.[12]

The hierarchical structure of some social species is also important for the automatic control of over-population. For instance, in a pride of lions access to food is determined by order in the hierarchy. When there is not enough food for everybody, least dominant individuals do not eat at all, which means the weakest cubs die off. This is advantageous for the future of a pride since strong fertile individuals are capable of making up for the population decline during the next period of abundance.

Memory also improves the long existing, genetically-controlled territorial instinct. A roaming group of higher-order animals can remember the boundaries of neighbouring unmarked territories by simply observing, from behind a tree on top of a small hill, a group of their neighbours enjoying their favorite blueberry patch. Then, they decide whether to fight for the new patch of blueberries or rather avoid risk of injury, even death, by respecting the current borderline.

It is already about 13.95 billion years after the big bang, or about 3.35 billion years after the origin of life on Earth, or about 50 million years BP. In lush forests of Africa, Eurasia and America, there are species of mammals that live in the forest canopy, where they are in relative safety out of the reach of predators prowling along the forest edges. Some of these species gradually developed into monkeys. They

---

[12] Robert Ardrey; *African Genesis*

became skillful acrobats, capable of moving quickly and easily through the environment full of branches.

During the several preceding millions of years, their front and hind limb inside fingers moved along their paws so that they now face the other four fingers. The inner fingers also became stronger and bigger and deserve separate names – thumb and big toe. As a result, each limb can now safely grip branches or other objects. The opposing thumbs are also very handy for peeling the inedible skins of different kinds of fruit.

The eyes of monkeys moved from the sides of their head to the front, where the flat face makes a suitable space for them. This enables a sharp, stereoscopic sight that permits much better judgment of distances between branches, and monkeys became even more agile than before.

The perfectly developed tail not only balances the monkey's body for nimble jumps from branch to branch, but it can also coil around the branches and provide additional grip and support.

At the ends of all digits, sensitive meaty tips appeared to crowd out previously strong and sharp claws. The claws became thin, flat nails. The claws – a natural weapon – are thus lost in favour of a better sense of touch. The meaty digit tips further improve their grip on branches. But monkeys do not remain without weapons. Four sharp canine teeth still protrude from their jaws.

Monkeys live in troops where juvenile monkeys learn many skills from their mothers and other members of the group. Finding varied food is an especially important skill. They eat fruit, nuts,

some roots, buds and leaves of plants.[13]  The
whole troop is available to help the mothers, so the
upbringing of youngsters becomes easier.  It is not
necessary to give birth only at springtime.  And
then a genetic mutation appears that changes the
timing of the fertile period of one female.  It is an
adaptive change, so it quickly spreads throughout
the population.  Female monkeys eventually become fertile about twelve times a year.  It enables
enlargement of the population, and monkeys expand into wider, bountiful territories.

Time has gone by and, in the course of about 20
million years, some kinds of monkeys gradually
became bigger, heavier and less agile than before.
It is now safer for them to limit jumps in favour of
swinging between branches from one arm to another.  They do not need a tail any more for this
kind of movement, so one of the genetic mutations
eventually removes it.  A new species thus comes
into existence and, at one time, they will be called
apes.  Together with monkeys they'll be called
primates.

As the apes search for many kinds of food
which ripen in different seasons, they have to
think about relations between weather, seasons
and the availability or lack of food in different
places at different times.  That's why every genetic
improvement of brain performance is an adaptive
change that quickly spreads throughout the whole
population of apes.

---

[13]  Desmond Morris; *The Naked Ape*

All species of apes face fewer enemies than smaller and weaker monkeys. That's why they dare to descend increasingly often from tree tops down to the forest undergrowth where they find new varieties of food. During about another 10 million years, several species of apes evolved into totally new species, very clever creatures that are capable of using suitable sticks, stones, perhaps also fragments of bones as tools for digging, peeling and crushing food.

Some ape species still remain mainly in the tree tops, but others stay mostly on the ground. That leads to a dissimilar development. For the apes in the tree tops, their body weight is supported by fingers as they grasp branches. The thumb grasps branches from below, and doesn't carry any weight, so it atrophies both in strength and size. The arms, backed up by a strong torso, become longer and stronger than before. As apes swing on branches, their rear limbs dangle under the torso, and they atrophy as did the thumb.

For the apes living mostly on the ground, the development is completely different. The use of tools held by front limbs, hands actually, encourages the ground apes to stand on their hind limbs increasingly often. The big toe moves back close to the fingers. The legs eventually completely lose the grasping ability. Shapes and positions of bones, joints, and muscles gradually change, until they allow walking and running on two legs only. This bipedal movement is slower than running on all four, but this disadvantage is balanced by the ability to tote tools while moving. All species with this ability will eventually be called humanoids.

Making and using tools is not easy. Suitable natural materials must be first selected, then adjusted to make tools. The using of tools requires development of new skills. All these activities present frequent and demanding exercises for the humanoid brain, which relentlessly increases in size and capability.

Each change in the genes of humanoids happens extremely infrequently. Even sogenetic changes, which are much more frequent, happen rarely. Sometimes hundreds, even thousands of generations do not witness any improvements in the organization and routines of their tribes. For them, life is changeless. They defend their territory against neighbours without hesitation, as their ancestors did. The genetic inheritance guides them blindly to fights for social ranking. They roam through the lush countryside and leave only a few traces after themselves and their campsites. Sometimes, the traces of their lives are turbulently covered with flood sediments or ashes of volcanic eruptions. Then, they rest covered – in petrified form – for millions of years as they mark faint outlines of primeval humanoid lives. One day, some of those humanoids will be named Sinantropus, Sahelantropus, Proconsul, Orrorin or Ramapithecus. But some will never get a name because their remnants will remain undiscovered.

The number of individuals in primate tribes varies from about ten to one hundred. Usually, a strong aggressive male is at the helm of a tribe. He is probably as smart as other members of his group, maybe a little smarter; that allows him to better direct the movement of his tribe through the countryside to find ripening food. This also may have been a reason why he became a leader, but his position was probably gained solely by his physical strength and skills in fighting for the dominant position of the group. In addition to preferential access to all fertile females, leaders have other advantages. Leaders of most primates approach food first in order to choose the best. After roaming the countryside for a whole day, nobody dares to settle down for the night until the leader is snuggled in the most comfortable spot. His strengths and aggressiveness is useful in infrequent territorial squabbles with neighbours or defensive actions against predators, but everybody participates in those, except mothers and youngsters. As long as the whole tribe keeps together, even large predators quickly learn to carefully respect them.

For some, but not all, species of apes and humanoids, the leader is the only male involved in reproduction. Gene passing to the next generation primarily by the strongest male insures the quality of the population. Partial participation of other males in reproduction ensures genetic diversity, but also the survival of the group if the leader happens to be infertile. Larger genetic diversity may not be necessarily useful, but sometimes a previously useless genetic combination could mean the salvation of the whole species after a later significant change of environment.

Just as with other species of primates, humanoids have well-developed territorial instinct. Quite naturally, they feel adversely to all neighbouring tribes. Different species fight for home territory with different enthusiasm. Some only yell and make lots of noise at the borders, but never engage in physical fights, and border changes are exceptional. Others take their border defense more seriously, though dangerous fights develop only rarely. The size of territories necessary for survival of humanoids is, in fact, considerable; the density of the population is therefore low, and skirmishes with neighbours are rare.

Even though many dangers lurk day and night for humanoids, the life of a tribe is quite easy most of the times, the life of a leader even easier due to the advantages ensured by his strength and status. Successful tribes that grow too large often split in two when a candidate for a leader attracts a few females to join him to find and defend a new territory against unfriendly tribes. Naturally, they remain friendly to the mother tribe. Some of the tribes, on the other hand, are unsuccessful, and they vanish. The number of tribes remains relatively stable. It is determined by the availability of food, which is relatively steady. Quite an idyllic way of life remains, in essence, exactly the same for generations. But after about 8 million years an enormous problem arises for all apes and humanoids.

# MAN

*COMPRESSED TIME.* Those who had enough patience and interest to follow this story to this point have arrived at the cradle of humankind. It is most likely somewhere in Africa around 10 million years ago. We know that we deal with an extremely long span of time, but can we really imagine how long a period of 10 million years is? For us, regular time is seconds, minutes, hours, days or years. We can still readily understand a decade if one is several decades old. We may even be able to perceive an average human lifespan, but the comparison of the human lifespan with the age of our beautiful planet Earth is beyond our imagination!

For better understanding, we need to compress the time a little bit in our mind's eye. We won't go all the way back to the big bang, nor to the origin of our planet, only to the origin of life on our planet. Pretend that the life begun at midnight and the present instant is noon. Therefore, the life so far has been evolving exactly twelve hours. In this compressed time, sexual reproduction started in the early morning at about 5:20. Nature created the first seashell about 10 o'clock. The first fish started to swim in the oceans about 20 minutes to 11 o'clock. Dinosaurs started to rule the world about 15 minutes past 11, and Archeopteryx – common ancestor of all birds – appeared about 11:30. Mammals started to replace the extinct dinosaurs about 20 minutes to 12. A few species of

apes and hominids reached the cradle of humankind approximately 2 minutes to 12 o'clock. Homo sapiens (the humanoid species we belong to)
left its cradle and started to build a civilization only about 20 seconds before high noon. Even a very
long human lifespan is such a short fraction of a
second in this compressed time-scale that it's
pointless to name this fraction.

INHOSPITABLE CRADLE OF HUMANKIND.
It is now about 12 million years BP. The cradle of
humankind doesn't look too amiable because of
serious and extensive droughts. Quickly disappearing African forests now receive only a third of
the precipitation they used to receive. And rain is
still waning. During these merciless droughts, the
amount of available food is shrinking. The majority of still available foraging food is in the forests,
but in the trees the ground apes and hominids can
hardly compete with more agile monkeys. Hard-
ship has struck all hominid species. Only the
strongest and the most resourceful ones will survive.

As time goes by, some hominids have already
become extinct, but several species are still dog-
gedly struggling with dry nature. Similarly, as
carnivorous bears learned at one time to eat plants
and roots in order to survive times of poor hunting, so do some hominids (foragers of fruits and
plants) now learn to hunt small prey. They add
meat to their formerly abundant, but now relentlessly disappearing, foraged vegetarian food

sources. It is not easy for them. When compared to sweet fruits, succulent sprouts and crunchy nuts, raw meat isn't as tasty, but it must suffice and their stomach luckily gets used to it. Difficulties with hunting are also hard to overcome. All prey that they, without claws and fangs, can subdue can run much faster than the hominids. Is their ingenuity and inventiveness going to be enough to enable their survival?

Yes, it is! About 4.5 million years BP, there are still several kinds of hominids successfully fighting against the dry, parched nature. The amount of rainfall has dropped again to about half of the rainfall of six or seven million years earlier. During this drought, organic remnants of dead animals and plants can't petrify, so hominids leave behind very few traces of themselves and their fight with arid nature. In spite of that, a few bones of hominids that died close to water are preserved as fossils and will form a little window into their struggle with the unbelievably long period of prehistoric drought. On the basis of these bones, one kind of hominid will be named Ardipithekus ramidus.[14]

This Ar. ramidus differs quite significantly from its bygone ape ancestors. His brain is larger, his hands are similar to apes' hands, but the thumbs are stronger. His posture is fully erect, and his legs are now longer than before, but not by much. His fangs (or canine teeth) are much

---

[14] Glenn C. Conroy; *Reconstructing Human Origins*

smaller. Natural selection allowed the loss of large fangs because their function was successfully taken over by weapons grasped in hand. The bodies of hominids don't show visible traces of natural weapons of their ancestors any longer. In spite of that they became successful hunters. A long time ago, one of them accidentally took hold of an antelope thigh bone and used it to finish a chase of some small prey. Gradually, their ingenuity enabled them to hunt for even bigger prey. The new hunting technique has been added to their two ancient fighting tasks: combat for dominant position in the group, and to the fight for territory for the group. And there is no reason why the hand armed with an accidentally-found object could not be used for all three types of combat.

After about another million years, a new hominid appears in the African savannahs. Perhaps he developed from Ar. ramidus, but most probably from another unknown hominid. He will be called Australopithecus afarensis, and his fangs are even smaller than the fangs of Ar. ramidus. The small fangs confirm that hominids already do not need natural weapons. Their fangs are now quite similar to incisors.

Most likely it is Au. afarensis that evolves – step by step, in the span of about 1.5 to 2 million years – into two branches: the later australopithecines and an altogether new hominid called Homo habilis, subsequently Homo ergaster, and finally Homo sapiens also called humankind, or mankind, or man.

BIRTH OF TECHNOLOGY. It's about 1.8 million years BP, and Homo ergaster – the only animal whose descendants will be able, at one time, to read this narrative – differs from other animals mainly by their erect posture and big brains that enable them to develop hunting techniques, although their bodies are poorly equipped for hunting.

Hunting was enabled by a set of innovations, among which we can count adapting and using natural objects as tools. The thigh bone of an antelope works as a bludgeon, a sharp-edged stone as a knife or scraper and teeth in the lower hyena jaw as a saw. Small bones can be sharpened for making holes into hides and other materials. A stone can be struck and broken to create a sharper edge. This edge can be improved further by skilful, repeated strikes. Production of stone tools and weapons is probably the first serious craft that the early man learns. Processing hide for a number of useful products, making wicker baskets, wooden tools and utensils, and utilizing other natural materials are slowly added to a growing number of crafts. Initially, there are only a few manufactured objects, and everybody makes only what is needed at the moment for personal use. While roaming through the countryside, people can carry only the most important tools in hand or small bags.

Development of craft skills is very slow. Youngsters learn by observing adults; this way craftsmanship is passed to a new generation.

Manufactured objects remain unchanged often for hundreds, thousands, perhaps even tens of thousands of years. Most improvements come into existence mostly by chance. Technological changes belong to sogens. As the thinking process of craftsmen improves by practice, some improvements may be the result of purposeful pondering. But any change, no matter how it arose, could be for better or worse. It is quite certain that many technological changes lead nowhere, the same way as mutation changes in nature do. The main principle of development is identical for nature, as well as society: changes arise to be tested, and then they survive or disappear. The difference is only in the number of changes and in the speed of development. Genetic changes arise without any selection; that's why many genetic mutations lead to idiotic changes that immediately perish. The pondering by man is not good enough to lead to success every time, but it is good enough to eliminate the most idiotic ideas from proceeding to a trial period. That's why the probability of success for tried technological changes is better than for genetic mutations. The ratio of adaptive to non-adaptive changes is higher for sogenes than for genes. A craftsman also tries new changes much more often than Nature. Development of technology is, therefore, faster than the evolution of nature, due to the elimination of senseless changes, and more frequent attempts for improvement.

*SOCIAL EVOLUTION.* The organization of the life of animals hardly ever changes. It does change for mankind, although almost imperceptibly. Similar to related hominids and apes, humans live in small, stable groups of about twenty to fifty individuals. According to their strength and skills, individuals rank themselves into a loose hierarchy with a dominant leader at its head. Hunting activities, forced on men and their ancestors by relentless droughts, require considerable cooperation among mainly the males of the tribe. Good protection and care of young – without which any group would perish – can be combined with food processing and with making many other products, and requires considerable cooperation among mostly the females. Roaming after ripening foraging food and migrating prey requires extensive collaboration of the whole tribe. Sick or injured individuals, pregnant women and women with infants need help and support from their mates, as well as the whole group. Life of the tribe is unthinkable without cooperation.

However, genetically conditioned fighting between individuals for dominance in the group or for a sexual partner, for food or comfortable sleeping spot, did not fully disappear yet, and probably never will. These kinds of friction can significantly disrupt cooperation.

During the cooperative execution of everyday duties, such as preparation for hunting excursions, processing of food, and securing camping grounds, and during well-deserved relaxation, many varied habits, practices, manners, morals, customs, traditions, rituals, celebrations and holidays slowly spring into existence. These elements, or sogenes, significantly influence the complicated workings of societies. Some of them later quietly disappear, others survive. Similar to technological mutation, new social changes may also be for better or worse. Some tribes have bad luck; their primary rituals are non-adaptive and do not contribute to the smooth working of the tribes. Malfunctioning tribes do not manage to replenish their population adequately, and their demise is then a harsh necessity. In contrast, morals and conventions of other tribes abundantly support cohesiveness, cooperation and successful survival of the tribes.

Celebrations and rituals facilitate the transition from childhood to productive engagement in the adult population. Pairing rituals of young men and women are especially significant, and they probably arise very early in the social life of humans. If habits and rituals govern and improve tribal life and make it generally easier, tribes grow and sometimes a few individuals splinter into a separate group. The new group is immediately organized due to habits, rituals and celebrations that everyone knows from the mother tribe. Most likely in summer (when the countryside is able to support a larger concentration of the population) all related, separate tribes gather to exchange goods and to facilitate the formation of reproductive pairs for those tribes where there are single young men or women. Sometimes they also pass newly-gained knowledge and experiences to each other.

A craftsman must rely on a proven routine to make the required number objects, but now and then he can try potential improvements. Changes in the organization of a tribe arise mostly by chance or by personal initiative of an influential individual. However, a substantial social, as well as technological, change doesn't happen even once in a generation. The success or failure of a technological change is usually evident soon. The significance of social changes is never obvious,[15] whether it improves society or contributes to its malfunction or even its demise. In some instances, an action of a tribe is determined by a leader or a group of elders that are guided by experience. However, it's unthinkable that a leader or elders would dare to institute a new ritual or to abolish an old one. Those arise or vanish entirely by chance, spontaneously, and very slowly.

Successful social rituals, whether they come to existence due to an obvious or for some completely unknown reasons, improve cohesiveness of the tribe and the dedication of the tribe's members to the whole tribe. That's why some individuals, in situations of impending danger, do not hesitate to accept, or even increase the risk for themselves, as long as it is favourable for their offspring, mate or even the whole group. This kind of behavior will

---

[15] Jan Keller; *Nedomyšlená společnost*

80

be known as altruism. Even primitive animals often behave in an altruistic manner; for them, altruism is encoded in their genes. As long as altruistic
individuals risk their safety (for example by loud
vocal warning against predatory beasts) they will
have higher mortality than others, therefore, they
should die out together with their altruistic gene.
But if altruistic behavior mostly helps blood-
related individuals in a group, and even if the al-
truists sometimes perish due to their altruistic act,
their genes survive in some members of the group.
The closer the group members are blood-related to
each other, the easier it is for a new altruistic gene
to spread through the group.[16]

A bee always dies when she stings. Yet, she
doesn't hesitate to use the sting to defend her sis-
ters. Such remarkable altruism has developed genetically just because all bees in one beehive are
genetically identical.

Ancestors of man have lived in small social
groups of mostly related individuals for several
million years already. That's why some social rituals that increase social cohesiveness only
strengthen already old, genetically-determined
altruism.

Tens of thousands, perhaps even hundreds of
thousands of years, pass by and, in some groups,
the dominant man (similarly as in many groups of
hominid ancestors) still has a preferential access to
all women for mating. It is not easy to achieve

[16] R. Boyd & J. B. Silk; *How Humans Evolved* (Hamilton's rule)

that. The sexual instinct incites everybody to mating, men all the time, women nearly all the time. In contrast to bygone ancestors, women are now fertile several days in every month, but it's impossible to tell when. Sexual activities are now much more frequent – practically constant – and attempts for copulation significantly disrupt much necessary cooperation in the tribes' activities. The evolutionary shift of sexual activities from public to private spaces[17] alleviates the problem, but not entirely. Preferential reproductive access of leaders to all women is quickly becoming a nonadaptive feature of tribal life.

In other tribes, the ritual coupling of maturing youngsters clearly indicates that the leader of the group does not insist on the sovereign mating right any more. Because the number of males is approximately equal to the number of females, each man now has regular access to at least one woman. This is a very desirable and pleasant situation since several biological mutations have increased sexual excitement. Undoubtedly, orgasm is the result of one of those mutations. Perhaps the increased sensitivity of naked skin, due to the loss of fur, contributes to sexual excitement, deepens the relationship of reproductive pairs, and leads males to the increased involvement in the care of their offspring. Over the course of time, the engagement of men in the care of their young is becoming ever more important. Human infants

---

[17] Jared Diamond; *The Third Chimpanzee*

are now born very small and helpless – somewhat undeveloped – since their brains and heads have increased in size, and the birthing passage from the uterus is no longer straight due to erect posture. In addition, the duration of upbringing gradually lengthens due to an ever increasing number of essential life skills that the young must learn. The care for young now requires more than the mother alone. Genetic, as well as sogenetic, mutations that increase the probability of survival of young undoubtedly belong among adaptive changes. Monogamy probably doesn't exist among those mutations yet, since deadly injuries may have incurred during hunting and contributed to a slight numerical advantage of women. But the majority of leaders already know that they not only have to allow but also encourage all men to have stable relations with at least one woman. What incentive would be there for male tribe members to cooperate with a tyrannical leader who prevents them to mate with females?

And so millennia goes by and it is already 6 million years since the split of hominids from apes happened and over 1 million years since the beginning of Homo species. Now, another landmark change is taking place in the life of man. It is about three quarters of a million years BP, perhaps a bit later, and man learns to use and control fire. By now, sogenetic changes have essential significance in humans' lives, and they considerably speed up the evolution of humans on our planet. Hominids have developed significantly bigger brains than other related animals, but otherwise hominids didn't differ from them too much until relatively recently. However, now people are emerging as extremely successful hunters and versatile foragers of fruits of land with very ingenious production of hunting weapons, tools, garments, decorative and ritual items and portable shelters for protection against the elements. Their technologies use many natural materials from stone to wood, grasses and bark, resourcefully combined with animal products, such as bone, hide, ligaments and teeth. A few dozen early signaling sounds – which originate in the vocal cords – develop into an articulate language with hundreds of words. As a result, better communication now improves organization of groups in all aspects. Upbringing and care of young is improving and lowers child mortality. Transfer of experience to new generations is more efficient than before. Various activities can be organized and better coordinated. Coordinated, complex hunting techniques, such as driving prey into traps or over natural cliffs, enable safer hunting of large prey in larger numbers.

Probably at some time in this period, a predominantly mental kind of work appeared. Thinking leads to observation of nature and attempting to explain various natural phenomena. However, that is impossible without a vision of supernatural beings. A shaman, or priest (who is best able to imagine those beings, or even communicate with them) is the first member of a tribe who doesn't have to participate in the manual chores of daily life. Special talents for observation also lead shamans to discoveries, gradually and by chance, of the medicinal properties of some plants.

The healing skills of shamans are based mainly on the purely empirical knowledge of the effect of natural plant substances on the human body. It's entirely different with their work of explaining natural phenomena and with their attempts to obtain favours and protection from supernatural beings. Here, shamans and their wards find themselves in a purely metaphysical, spiritual domain. They strongly believe that by proper deeds, which include appropriate gifts or sacrifices, they can win supernatural beings over to their side, and induce them to restrict storms, excess of rain or, by contrast, the length of drought, or to cause an abundance of hunting prey. They imagine supernatural beings as ephemeral ghosts or goddesses and gods in various animal, human or combined forms. Some ghosts represent dead ancestors that also must be worshiped, lest they help, or at least not harm, their descendants.

All people believe that each individual is born under the influence of some kind of animal essence or a ghost. Shamans are best able to determine under what influence each child is born, and they assign them a specific amulet called a totem that represents this influence. A boy with the lion totem will most likely become a proficient hunter, while a girl with a rabbit totem will probably be a very good messenger, until, of course, she starts to bear children.

People feel a particularly great respect towards gods, and they fear gods' terrifying almightiness. But they still do not have even an inkling about the unifying, all explaining essence of Nature. Rather, they divide natural phenomena into different domains that are under control of different gods. Miraculous phenomena that belong to the sphere of birth of new life and constant natural renewal are nearly always considered under the power of goddesses, since only women (as with female animals) are apparently chosen by goddesses to give birth.

The perceived ability of supernatural beings to harm or help people is almost always on the mind of early man. The shaman leads the tribe to behavior that is approved by gods, and helps people to limit their fear of the supernatural beings – particularly when natural catastrophic phenomena strike. That is why he or she belongs among the most important and influential members of all tribes. A shaman is convinced of his abilities to discern what gods prefer, to hear their instructions, and to interpret them correctly. This talent, often strengthen by extraordinary, even schizophrenic mental anomalies, together with thorough apprenticeship with an elder shaman, gives him an absolutely unique position in the life of his tribe. Not even leaders of tribes dare to interfere in the shamans' domain. Perhaps eventually, some shrewd individuals learn to act as if they understand gods' instructions just to obtain privileged position in the tribe.

Individual differences in competence and strength lead to the first significant, but still quite simple, specialization and division of work. Able and swift men devote themselves predominantly to hunting. When a dexterous young man limps heavily, due to an unfortunate injury, he will probably concentrate on production of weapons or other items. Eventually, nobody can match his skills, so he remains the only person of his tribe involved in his new specialization. Expert craftsmen do not make things only for themselves but for the whole tribe. When it comes to hunting weapons, those belong to hunters, but their catch belongs to the whole tribe.

Even the production of basic garments and preparation of food has a communal character. It is up to the leaders to see to the distribution of the communal goods to members of the tribe according to their needs. But the production of decorative and other not too numerous luxury items slowly leads to barter trade and the concept of personal property. As yet embryonic personal property forces its way to existence first in a small pouch with personal amulets, which are, in a superstitious society, important for everybody for obtaining protection of gods against danger. Trade in goods between related tribes takes place at summer tribal camps arranged by leaders and elders.

Many other features of the tribes' social life shape up in this period, and most of them are based on sogenetic mutations that have worked about 1.5 million years already, in comparison to genetic mutations that have worked 3.5 billion years. Social evolution is a very slow process but much faster than genetic evolution. Because of their inventiveness, humans now populate a larger area than most other species. They live in small social groups. When population grows, the size of groups remains relatively unchanged, but the number of groups is growing as they are spreading to ever larger area. Many generations of man lived over the last 1.5 million years. A very large number of sogenetic changes arose, were tested, and accumulated during this time in a large number of more or less isolated tribes.

The success of random changes depends on a huge number of trials leading to many, many errors and few improvements. For genetic changes, this huge number is facilitated mainly by enormous lengths of time. For sogenetic changes, time is also important, but the huge number of trials is mostly the result of the size of population living in a large number of groups.[18]

So, there is now a little less than half a million years BP and man can already use, maintain and control fire. Together with the improving production of garments, fire enables humans to spread far and wide, to even very cold regions. With the

---

[18] See page 106, *Appendix 3* for more information

enlarged occupied territory, the human population also increases, and the probability of the birth of adaptive sogenetic changes within tribes increases as well. Besides warmth and night protection against beasts of prey, fire also finally enables people to improve the taste of their relatively new food of animal origin. At the same time, it enables the addition to the already very diverse diet of further plants and roots that are digestible only in a cooked state. Of course, all these culinary inventions emerge very slowly over the course of tens of thousands of years.

The troubles with droughts lasting 11 million years disappeared as recently as half a million years ago, and the amount of moisture then started to fluctuate significantly. In the course of these years, precipitation first increased about sevenfold, then it dropped to less than half of the highest level, only to rise even more than the first time. Temperature fluctuates together with moisture in a mutual relation. The higher the temperature goes, the more water evaporates into clouds. More clouds screen out more solar energy and the temperature can drop down again. It can, but won't necessarily, since many other influences contribute to the very complicated system of the planet's climate.

Volcanoes often spew out a huge amount of fine ash that floats in the atmosphere for years. Sometimes there is so much ash in the air that it screens out a significant part of solar energy. Lightning sometimes ignites forest fires that burn often for months. Gases from burning wood, including carbon dioxide ($CO_2$), change the composition of the atmosphere, and it becomes partially reflective. Then, some of the solar rays that bounced off the Earth's surface do not escape out into cosmic space, as they reflect back down to Earth. At one time, this occurrence will be called the greenhouse effect; of course, it raises the temperature.

All green plants need $CO_2$ for their metabolism. Extensive new forests consume so much carbon dioxide that they lower its amount in the atmosphere and thus reduce the greenhouse effect. At the floor of old growth forests, rotting vegetation generates $CO_2$, and thus increases the greenhouse effect again. Plant life decaying under the pressure of deposited earth is transformed into crude oil or natural gas. The Earth's crust saturated by natural gas is often flooded by deep oceans. Under the pressure of soil and the chilling ocean depth, natural gas freezes into a solid state. At some places, underwater volcanic eruptions light up this frozen gas and thus release an enormous amount of $CO_2$ into the atmosphere.

The ocean waters warm up by different degrees at various spots on Earth, especially at the surface and at the bottoms of shallow seas. Warm water is lighter and rises up, cold water descends down. Together with winds, the uneven water temperature causes strong ocean currents. Warm currents transport heat for hundreds of kilometers away from the equator – where solar radiation is the strongest – and warm up distant lands. When a warm period lasts long enough, it can melt large amount of polar and high mountain glaciers. The melting of fresh-water ice reduces the concentration of salt in the ocean's water. Fresh water is lighter, and as it mixes with heavier salt water it may start new ocean currents or stop old ones. These changing currents could stop the transfer of heat away from the equator, and distant lands could cool down rapidly.

Only Nature knows precisely what multitude of influences participate in the creation of weather. One thing is certain: climate changes cause the first ice age, at about 400,000 years BP. Glaciers spread far into many regions. The ice annihilates plants, many animals and some tribes of man, and other tribes are pushed out from many places of the Earth. Fire gives them life-giving warmth and time to migrate to warmer regions. Over the course of time, and with rising temperatures, glaciers start to melt and retreat again to the polar regions and summits of the highest mountains only, but later, they spread out several times more.

Fire also opens new and far-reaching technological possibilities. Easier sharpening of sticks and small bones is one of the first utilizations of fire. A surprisingly hard lump of clay found in a cold fireplace, which previously fell in the fire by chance as a shapeless lump of moist clay, or noticed at the site of previous forest fire, leads to ceramic crafts. Perhaps a little bit earlier, blackened embers from a cold fireplace provided black col-

our for the first visual artists. These black embers
consist mostly of carbon atoms. This form of carbon is called graphite; it is not by accident that
drawn artistic expressions will later be called the
graphic arts. Somewhat realistic drawings on rock
walls of caves, which are used as dwelling shelters
or ritual spaces, give evidence of a hugely significant skill: abstract thinking and usage of symbols
for various objectives. Mural drawings are at the
beginning of an important development of graphic, later on written expression – the way of communication that can bridge over time.

Abstract thinking of man causes a major move
forward in the development of human societies,[19]
but only exceptional personalities can devote
themselves to it at this time. Very few individuals
create any kind of graphic expression, whether a
mural in a ritual cave, or a few notches on a stick
to record a number. Few people can count to
more than ten. Only shamans are beginning to
discover methods of counting a large number of
animals or objects. Making any kind of record is
still beyond the ability of most people. When
people barter small objects, it is relatively easy to
achieve an immediate settlement, but it is much
more difficult with larger objects. Unsettled
trades of items or services must stay in a person's
memory, and can influence the status of debtors

---

[19] Jared Diamond; *The Third Chimpanzee*

and creditors. The owing of debts lowers status, paying debts off increases status.[20] Naturally, unsettled trades rendered to the leader or to the whole tribe increase status the most.

The number and complexity of specialized crafts for production of utility objects grows, and there are so many crafts already that few tribes master all of them. That leads to barter trade among related tribes. But trading with unfriendly tribes is not easy. Territorial instinct has influenced ancestors of man for millions of years already. Hostility, even hatred of the unrelated tribes is strongly rooted. On top of that, due to the size of territories required for survival, it's not easy to have any contact with unfriendly tribes, much less to trade with them. Contact with unfriendly tribes is likely limited to infrequent border skirmishes.

Quite possibly, especially with tribes where the leader still insists on priority access to all females, temporary proximity to an unfriendly tribe is exploited by a group of eager young men for ambushing and raping women that gather blueberries in a group at a remote clearing.[21] At one time much later, rape will be considered as something very ugly, but to the prehistoric youngsters it doesn't seem that their behavior is wrong at all. In their minds, the enemy tribes do not deserve any regard or respect; on the contrary, they are consid-

---

[20] Jean M. Auel; *The Land of Painted Caves*
[21] Jean M. Auel; *The Mammoth Hunters*

ered as something different, untrustworthy, even inhuman.

A similar group of young women can be surprised by a small group of young hunters from an enemy tribe, in which sexual fidelity is already a thousand-year long tradition. Two men of this tribe may have just lost their mates in an unfortunate natural disaster. The tribe is at the edge of extinction since it has less than ten adult members now and very few women. The hunters kidnap two women for the widowers and become exposed to a substantial risk of revenge. Even though they think of them as inferior, they will treat the captured women with respect and hope that they will help them keep their tribe viable. Another tribe, in a similarly grave situation, chooses a less risky solution, trying to buy surplus women from neighbours. But it will not be easy to contact distrustful enemies, and they will have to pay with a large number of their best weapons and other highly-valued objects. Trading with enemies is difficult; it can be done only in exceptional situations.

Hunting techniques keep improving, and men now skillfully throw weapons, such as javelins and rocks together with new devices, such as slings, slingshots and javelin throwers. Perhaps it is exactly the throwing technique that helps people in the struggle with other hominids for adequate territory. The penultimate hominid, who will be called Neanderthal, disappears from Earth's surface about 30,000 years BP. Man re-

mains on Earth as the only animal with a large
brain and fully erect posture. His ingenuity is still
without limits.

An entirely new sogen appears in a mere few
thousand years. While roaming through the countryside, a member of a tribe, probably a young
girl, finds small wolf pups without parents. Driv-
en by the motherly instinct, she takes the pups into her care. A similar situation occurred many
times before with small baby goats and baby
sheep, whose parents become game food for the
tribe. But those young animals are too consumed,
sooner or later, at times of lesser abundance. Wolf
pups, however, are not part of the human diet, so
they become fully grown in human care. Surprisingly, the tame wolves turn out to be capable as-
sistants in human hunting activities. Perhaps after
another twenty or thirty years, or after few centu-
ries, the experience with breeding dogs, the descendants of wolves, leads to breeding sheep and
goats. It soon becomes apparent that pastoral
farming is an revolutionary social change. It allows better control over the meat diet and makes
hunting activities less important, while food sup-
plies improve and the human population grows.
In addition, people soon learn to use milk of ewes
and mother goats as human food, when the animal young are still-borne. People also soon learn
how to keep milk flowing even after the young are
weaned.

Pastoral farming significantly complements
other sources of food. About 12,000 years BP, pas-

toral farming overshadows other ways of subsistence in some tribes, but it doesn't otherwise change their life too much. They still have to roam the countryside constantly in harmony with natural cycles. Incessantly, they have to drive their herds towards new pastures largely in high elevations where trees do not grow. Mountain meadows are grazed down often in just one day. Before winter, they must come down to lower, warmer elevations, where winter vegetation can maintain reduced herds, perhaps also with the help of sickle-cut and stored grass. When reducing herds, people first consume the males. Only the best ram and billy goat are saved to impregnate the females. Lastly, the infertile and weak sheep and goats are consumed, but pregnant females must be maintained to give birth to the new herd next spring.

In the difficult task of maintaining herds during cold winter months, one trick of Nature comes to man's assistance. Grasses reproduce by both a vegetative and a sexual way. Two different kinds of grasses fertilize each other and produce a third, unique grass. In nature, this happens quite often. The resulting product of this process will be called crossbreed or hybrid. Hybrids are usually sexually infertile, but not always. This time, the new grass hybrid is fertile, and it has larger seed clusters (the ears) than both parents. The seeds in the ear are also larger[22]. People harvest this grass

[22] Jacob Bronowsky; *The Ascent of Man*

(called emmer) since its full ears are more nourish-
ing for the winter herds. The lost and much later
found sickles (composed of stone blades and
wood or bone handles) will witness this activity
for us at one time. At the slopes of wintering valleys, remaining ripe ears of emmer fall apart, and
the seeds – dispersed by winds – help to ensure
the new harvest next year.

   Now, it will take just a few thousand years till
another most far-reaching change will start a
completely new era in the life of man. But there
will be a different narrative about that.

❖

Thank you for going, with the author, through about 14 billion years of nature's development. Please consider now writing a brief review of this book at Amazon or Goodreads.

Also please consider answering 3 'multiple choice' questions here : www.surveymonkey.com/s/VYMG5MV. The survey-monkey website will not show the identity of responders to the author.

If you like, email a brief comment to the author. He will be grateful for each response, especially critical or even one word comment: parpbook@gmail.com

# APPENDIX 1

SPONTANEOUS GENESIS OF ORDER FROM CHAOS. To study the net of molecules in the primeval soup of life would be difficult because it's too large and hugely complicated. For easier understanding, we need a simplification. Researcher and author, professor Kauffman[23] devised and studied one such simplification.

He replaced the complex net of molecules with a simple net of blinking bulbs. We will first look at a net of three bulbs. Any smaller net would be hard to make (Pic.8). At each predetermined point in time, each bulb will come on or go off according to the state of the neighbouring bulbs in the preceding moment. First, we'll determine the rules that the bulbs will have to follow.

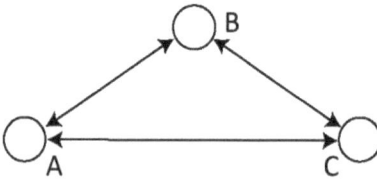

Pic.8

Bulb A will come on when both bulbs B and C were on in the preceding moment (Boolean logic AND). In all other possible situations, bulb A will go off. Bulbs B and C will come on when one or the other neighbouring bulb is on in the preceding moment (Boolean logic OR). In all other possible situations, bulbs B & C must go off. The input or signal from two bulbs has only four variations.

---

[23] Stuart A. Kauffman; *At Home in Universe*

The reactions of all bulbs to all inputs are recorded in table 1. Naturally, 0 stands for off and 1
stands for on.

| Reaction of A (by logic AND) | | Reaction of B (by logic OR) | | Reaction of C (by logic OR) | |
|---|---|---|---|---|---|
| Input | Reaction | Input | Reaction | Input | Reaction |
| B C | A | A C | B | A B | C |
| 0 0 | 0 | 0 0 | 0 | 0 0 | 0 |
| 0 1 | 0 | 0 1 | 1 | 0 1 | 1 |
| 1 0 | 0 | 1 0 | 1 | 1 0 | 1 |
| 1 1 | 1 | 1 1 | 1 | 1 1 | 1 |

Table 1

In accordance with Table 1, we now record
what will be the state of the whole net after each
variation of the input (Table 2). There are only 8
variations of input in this small net (2 states, 3
bulbs: $2^3 = 8$).

| Input | Reaction | Input | Reaction |
|---|---|---|---|
| A B C | A B C | A B C | A B C |
| 0 0 0 | 0 0 0 | 1 0 0 | 0 1 1 |
| 0 0 1 | 0 1 0 | 1 0 1 | 0 1 1 |
| 0 1 0 | 0 0 1 | 1 1 0 | 0 1 1 |
| 0 1 1 | 1 1 1 | 1 1 1 | 1 1 1 |

Table 2

Now, we'll start the net blinking and watch it.
There are 8 possible starts overall (Table 3).

Rigid stillness will set in six times, once with all
bulbs permanently off, and five times with all
bulbs constantly glowing (starts 1, 4, 5, 6, 7 and 8).
State 111 can be called an attractor, since it

happens most often. In two steps after two starts, the net will walk into a perfect order and repeatedly flicker between two states (starts 2 and 3). Perhaps, it is not surprising for such a small net that the net will never blink chaotically.

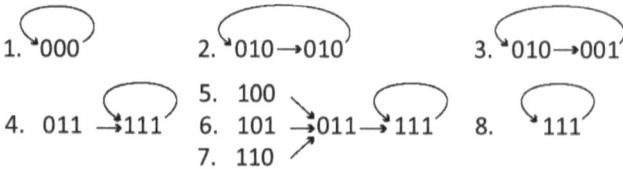

1. ↶000↷   2. ↶010→010↷   3. ↶010→001↷

4. 011 →111   5. 100 ↘   8. ↶111↷   Table 3
   6. 101 →011→111
   7. 110 ↗

But what will happen if we considerably enlarge the net, say to 1,000 bulbs? This number of bulbs can go through quite an astronomical number of blinking combinations ($2^N$, N=1,000). Should this net start going through all its possible states right at the origin of the universe, and should the interval between states be only one trillionth of a second, this net will not be finished with its task today.

However, we won't be content with 1,000 bulbs; we will promptly take 100,000 bulbs (N=100,000), so that the number of bulbs will better correspond with the number of organic molecules in the soup of life. Each bulb will be connected with only two neighbours (K=2). Now, could this unimaginably huge net ever jump into perfect order? Let's turn to the computer of professor Kauffman for an answer.

This net will start with chaotic blinking of almost all bulbs, then some bulbs will gradually become either switched on or switched off, and the rest of the bulbs will rather quickly stabilize into neat cycling among only 317 states! The cycling among these 317 states is an attractor that will be reached from several starts. Wow! Who has only a rough idea of the huge size of the net and of a potential possibility of utterly chaotic and eternal blinking must be stunned by the genesis of such an exact order.

The molecular mess in the soup of life is more complex by far than the net of bulbs where N=100,000 and K=2. N is large enough, but K is not, since each type of molecule can be influenced by more than two neighbouring molecules. That's why we now connect each bulb with 4 neighbours (K=4). When we start up the net now, we'll see mostly chaos or rigid stillness. But not inevitably.

From four neighbours, each bulb can get 16 different inputs (2 states, 4 neighbours: $2^4$=16). For a bulb, there are only 2 possible reactions to each input: '0' or '1'. Depending on the chosen logic, the reaction to 16 inputs could be '0' for all inputs, or '1' for all inputs, or a combination of '0' and '1' (Table 4). Now we'll start changing the logic of the bulbs' reaction, and hope that an orderly blinking will result at some point. For each set of reactions, we'll divide the number of reactions '1' by 16 – that is by the total of reactions in one set. The ratio 'P' obtained like this will move from 0 (P= 0/16= 0) to 1 (P= 16/16= 1).

This complex net – alike a simple net – will remain in rigid stillness when all of the bulbs react to all inputs by '0' or '1'. When 'P' is near 0 or 1, the net will start blinking, and it will bog down

soon into rigid stillness. If we try P=0.5 the net
will start blinking chaotically, and chaos will last
forever.

Reaction of typical bulb

| Input | Reaction | Input | Reaction | | |
|---|---|---|---|---|---|
| 1. 0000 | 0 | 9. 1000 | 1 | | |
| 2. 0001 | 0 | 10. 1001 | 1 | | |
| 3. 0010 | 0 | 11. 1010 | 1 | | |
| 4. 0011 | 0 | 12. 1011 | 1 | | |
| 5. 0100 | 0 | 13. 1100 | 1 | | |
| 6. 0101 | 0 | 14. 1101 | 1 | | |
| 7. 0110 | 0 | 15. 1110 | 1 | | |
| 8. 0111 | 0 | 16. 1111 | 1 | P= 8/16= 0.5 | Table 4 |

Now let's try raising the value of P gradually.
When P is somewhere close to 0.75, the net walks
into perfect order. Without organizational en-
deavour, without effort, the order sets in. At the
boundary between chaos and stillness there lies
the order.

Naturally, blinking bulbs do not indicate how
life originated, no matter how orderly they are, no
matter that they are in a net as vast as a 100,000
bulbs. But they enable us to sense that – in highly
complex, natural systems – crystallization of
spontaneous order is possible. Such spontaneous
order can happen because the chemical laws, the
laws of complexity and the principles of evolution
allow it.

*Appendix 1 relates to page 37.*

# APPENDIX 2

*FITNESS LANDSCAPE.* The size of fitness landscape grows quickly with the number of genes. The more genes an organism has, the more possibilities there are for evolutionary changes, and higher probability for the adaptive changes. Should an organism have only 3 genes, each with 2 alleles, it would have only 8 possible variations or genotypes (2 states, 3 genes: $2^3$=8 genotypes). Each additional gene doubles the size of the fitness landscape. For a species with 4 genes, each with 2 alleles, the number of possible genotypes is 16 ($2^4$=16 genotypes).

As an example of a simplified explanation of fitness landscape, we choose a creature called trid. Trid is an unreal (and totally impossible) animal with only 3 genes. Each gene determines one characteristic of trid, and potentially, each characterristic can have only 2 states (alleles) such as green or brown, smooth or rough, or something similar. Over time, trid can mutate to 8 variations or genotypes (2 alleles, 3 genes: $2^3$=8 genotypes). The genotype that has all 3 genes in the first state will be designated as 000. Number 111 indicates 3 genes in the second state.

Now, we will construct the fitness landscape of trid. Let's take 8 buttons and scatter them over the plane of zero fitness. Each button represents one possible genotype of trid. Now, we will assign a random value of fitness, from 1 to 10, to each genotype, and raise each representative button to a corresponding fitness level above the plane of zero fitness. (Pic.9)

In one step, each genotype can mutate into 1 of 3 neighbouring genotypes. We will depict that by drawing arrows between the neighbouring genotypes. The arrows always have to point to the genotype with higher fitness. The arrows now represent the mutation paths.

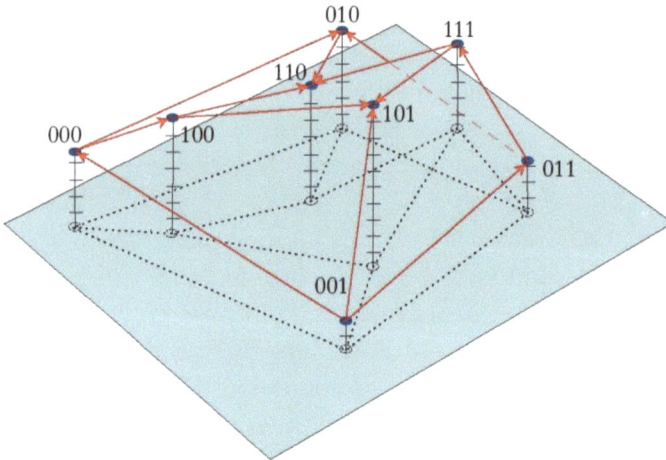

Pic.9

Now, it only remains to add the surface of the fitness landscape (Pic.10). The path from the point 011 to the point 010 leads through a short underground tunnel. The very first trid must have had relatively low fitness; let's say that it was in the point 000. If the next mutation leads to the point 001, the trid population will stay in the point 000. That is because the individuals 001 are less fit, and they won't be able to compete for resources with the more fit individuals 000. One of the paths trid can follow is 000 – 010 – 110. In the point 110, trid will get permanently stuck, since all directions path from this points lead to lower points.

Only the 000 - 100 – 101 leads from the point 000
to the highest possible fitness for trid in the unchanged environment. However, the environment
can change.

Let's say that trid 110 is green and potential trid
111 would be brown. It's the green colour that
provides trid 110 with the higher fitness, because

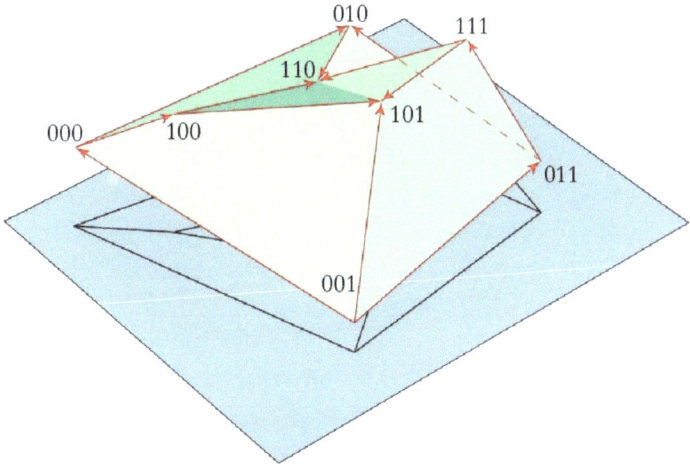

Pic.10

predators that eat trids do not see them very well in
the green environment in which trids live. But
now, the country became brown, due to a spell of
drought. The mutation into the state 111 became
suddenly advantageous in the new brown
environment. The point 110 dropped lower while
the point 111 was raised up. The lucky individuals
with the brown mutation now survive a lot better
than the green trids and the next mutation can
propel brown trids to the highest elevation of the
trid fitness landscape – the point 101.

*Appendix 2 relates to page 47.*

## APPENDIX 3

SOCIAL EVOLUTION. The evolution of nature is driven by a huge number of random changes and the process of natural selection. Social evolution also depends on many changes that come randomly in much faster succession, over a much shorter time. The following simple computations clarify some of the differences between genetic and sogenetic evolution.

The average human life span is, in the prehistoric times, about 25 to 30 years. Childbearing begins with sexual maturity, which is completed between the ages of 11 to 13. One generation span is, therefore, about 15 years long. In the first 1.5 million years of human social evolution, there were about 100,000 generations (1,500,000/15).

With their thin fur and later virtually hairless skin, humans were originally limited to warm climates, but with garment production and control of fire, the human population is spreading to ever larger areas. Definitely, people are already in the territory of Africa, west Asia, India, Java, east China and west Europe. Let's say that the total population of humans is now 700,000. The size of a typical social group is about 20 to 100, perhaps 30 on average. The total number of tribes is, therefore, about 23,000 (700,000/30). During the last 1.5 million years there successively were 2,300,000,000 separately living generations (100,000 generations times 23,000 tribes). The huge number of small, nearly imperceptible sogenetic changes could have originated, been refined and accumulated in this huge number of generations. For an individual, these changes are almost imperceptible, but they take place at an incredibly fast pace, when compared to the speed of biological, genetic evolution. In compressed time, in which *life evolution lasts 12 hours* and the present is high noon, hominids originated about 1 minute before noon, and *1.5 million years* – from the origin of hominids to the regular use of fire by man – are represented by only about *17 seconds* of this compressed time.

The success of random changes depends on a huge number of trials leading to many errors per improvement. For genetic changes, this huge number is enabled mainly by the enormous length of time (3.5 billion years). Time is also important for sogenetic changes, but a huge number of trials is given mostly by the huge number of separately living generations over time (2.3 billion generations). The large number of social groups, in combination with large brain and information exchange between groups, enables sogenetic evolution to accumulate a large number of more or less random changes that lead to the development of complicated, well functioning systems. The length of time is important for the social evolution of man, but its importance is decreasing and it is quite negligible in comparison to the length of time of biological evolution *(17 sec. against 12 hrs).*

*Appendix 3 relates to page 87.*

❖

www.ingramcontent.com/pod-product-compliance
Lightning Source LLC
Chambersburg PA
CBHW041228270326
41935CB00002B/7